Peleliu Progress

Other Tactics/Intel. Manuals from Posterity Press:

The Iwo Alamo
Super-Squad: The Now Missing Component
North Korea's Hidden Assets
Guardian Joe: How Less Force Helps the Warrior
Sinoland: The Subversion of Freedom's Bastion
Afrique: A Warning for America
Strategic Rifleman: Key to More Moral Warfare
Gung Ho: The Corps' Most Progressive Tradition
Global Warrior: Averting WWIII
Expeditionary Eagles: Outmaneuvering the Taliban
Homeland Siege: Tactics for Police and Military
Tequila Junction: 4th-Generation Counterinsurgency
Dragon Days: Time for "Unconventional" Tactics
Terrorist Trail: Backtracking the Foreign Fighter
Militant Tricks: Battlefield Ruses of the Islamic Militant
Tactics of the Crescent Moon: Militant Muslim Combat Methods
The Tiger's Way: A U.S. Private's Best Chance for Survival
Phantom Soldier: The Enemy's Answer to U.S. Firepower
One More Bridge to Cross: Lowering the Cost of War
The Last Hundred Yards: The NCO's Contribution to Warfare

PELELIU PROGRESS

ILLUSTRATED EDITION

H. JOHN POOLE

FOREWORD BY

GEN. ANTHONY C. ZINNI USMC (RET.)

POSTERITY
PRESS®

Published by Posterity Press
P.O. Box 5360, Emerald Isle, NC 28594
(www.posteritypress.org)

Cataloging-in-Publication Data
Poole, H. John, 1943-
Peleliu Progress
Includes bibliography and index.
1. Infantry drill and tactics.
2. Military art and science.
3. Military history.
I. Title. ISBN: 978-1-7354530-1-9 2021 355'.42
Library of Congress Control Number: 2021904969

Edited by Dr. Mary Beth Poole

First printing, United States of America, June 2021

For all present-day American warriors who realize that much of what their predecessors learned was never incorporated into their organizational wisdom.

Within the cover art is "The 2000 Yard Stare" by Thomas Lea, a *Life Magazine* correspondent with 1st Marine Division at Peleliu.

As at Guadalcanal, the U.S. Army would be called upon to consolidate all U.S. Marine Corps' gains on Peleliu. That task would fall to the same 81st Infantry Division units that served as Peleliu invasion force reserve. Some had just seized Angaur Island seven miles to the south. Due to the labyrinth of sub-surface passageways on Peleliu, that consolidation process would take an additional month and a half. The hardest part would come at the heart of the Japanese defense complex atop the southern end of Umurbrogol Mountain.

The cover art face may be that of someone with combat fatigue, but it does not mirror the demeanor of the average Marine infantryman on Peleliu. The self-propelled artillery piece in the background appears to be a U.S. Army M8. All Marine artillery was of the towed variety. The 321st and 323rd Infantry Regiments would need that extra firepower to clean out the final enemy pocket created by the 5th and 7th Marines.

Contents

ILLUSTRATIONS

Maps:

Figures:

Foreword

Unlike the European Theater in World War II, there was no single Pacific Theater or Command. The Pacific was divided into several commands. The two major commands were General MacArthur's Southwest Pacific Command and Admiral Chester Nimitz's Pacific Oceans Area Command. Each of these commanders proposed their own strategies to close the ring around the Japanese Empire with President Roosevelt choosing not to opt for one course, but to allow both to take their proposed routes toward the Japanese main islands. MacArthur's plan was to come at Japan from the south through New Guinea and the Philippines, while Nimitz planned a central Pacific drive through the Marianas and Iwo Jima. Each drive would culminate with either the taking of Okinawa or Formosa as the final jumping off position before invading Japan proper.

Between these two routes to Japan lay several island chains occupied by Japanese forces that could threaten the flanks of both planned campaigns. One of the chains was Palau with the island of Peleliu in the chain containing a Japanese air and naval base. A debate ensued as to which islands posed significant threats, which islands could be bypassed, which could be reduced without invasion, and which had to be taken through potentially high casualty amphibious invasion. It was decided that Peleliu should be taken and what resulted was the highest casualty producing battle of World War II. What also resulted was a highly controversial look back at Operation Stalemate, as the Peleliu invasion was titled.

John Poole has once again analyzed and dissected a historical military event and drawn out critical lessons for us today. He begins by making an insightful comparison to Japan's Imperialist designs and China's current expansion in the Pacific. As John points out, the so-called Japanese Greater East Asia Co-Prosperity Sphere looks remarkably like China's current military and influence stretch into neighboring Pacific regions. The same Pacific island nations and chains that became contested areas in World War II are once again seen as potential areas of conflict. The strategic implications are similar in many ways to those we faced as Japan expanded its power; however, there are differences in the modern military capabilities that would be involved.

One of the most important lessons that is evident in John's analysis is the lack of critical intelligence on the enemy and environment. Japanese defensive tactics, changes in counter amphibious operations, and extensive defensive construction were not anticipated. As a result, this casualty heavy operation caused the First Marine Division to suffer one third of its force to be listed as killed or wounded. Environmental intelligence requirements that were learned the hard way on island battles such as Tawara led Admiral Kelly Turner, Nimitz's brilliant commander of his amphibious forces, to realize his shortfall in intelligence on tides and beach obstacles, both natural and man-made, and created the Underwater Demolition Teams structure to attempt to gain more intelligence on these critical aspects of launching a successful amphibious landing.

John Poole is adept at drawing out strategic and operational lessons; however, he is especially knowledgeable and experienced in two areas—small units tactics and small unit leadership. He understands infantry and close battle. He captures the Battle of Peleliu from this perspective, then poses several key questions. Could we face this kind of fight again? Do our current forces possess the leadership development, training, and tactics to succeed in this kind of fight?

These are serious questions that our Services must answer. There are critical implications regarding our training, education, leader development, equipping, and structure when answering these questions. Have our combat arms units, particularly the infantry units, become overloaded, road bound, operating-base tethered, distracted by social changes, over-reliant on technology, and prevented from experiencing stressful training to the extent that facing a battle like Peleliu would result in a disastrous outcome?

GEN. ANTHONY C. ZINNI USMC (RET.)
FORMER HEAD OF CENTCOM

PREFACE

All members of the U.S. Marine Corps and its larger plus better equipped sister service—the U.S. Army—have heard of the Battle of Peleliu. It was the first of four island campaigns late in WWII where the Japanese used a new defense strategy—not trying right away to repel invaders. The other three applications were at Luzon, Iwo Jima, and Okinawa.

Foremost in the modern-day Government Issue's (GI's) recollection of Peleliu are that a lot of people got hurt and that a senior naval officer suggested the island didn't need to be taken.

This book has been written for three reasons: (1) to detail the methods of the enemy's new strategy; (2) to determine the tactical innovations with which the Americans countered those methods; and (3) to see whether the Peleliu struggle had served a useful purpose.

Tactical hindsight can be misleading. Other variables often determine who does well on the field honor. All U.S. service personnel who fought for Peleliu were exceptionally brave. They may have won through never hesitating to move forward.

Lt.Col. H. John Poole USMC (Ret.)
career infantryman and
former Fleet Marine Force Gy.Sgt.

Acknowledgments

Thanks be to God for everything useful. May the sacred war dead of Peleliu now rest in peace knowing their supreme sacrifice has given America the warfighting expertise to maintain worldwide freedom.

Part One

The Effect of History on Future Events

"It is right to learn even from the enemy."
— Roman poet Ovid at the time of Christ

(Source: Attributed to Ovid.)

1 The Chinese Communist Co-Prosperity Sphere

- What was the Greater East Asia Co-Prosperity Sphere?
- Which islands in the Western Pacific did Japan occupy?

China has long considered itself center of the civilized world.

(Source: With permission of IISH/Stefan R. Landsberger Collections at Leiden Univ. in the Netherlands, ©, from ChinaPosters.net image designator "E13-716.jpg")

China Claims Ownership to the Entire South China Sea

For decades, the People's Republic of China (PRC) has gradually consolidated its hold over the Western Pacific—to include the 1.35 million square miles historically called the South China Sea. (See Maps 1.1 and 1.2.) That 1947-born nation's justification for this "power grab" is that no one disputed Chinese sovereignty over the region prior to 1930.[1] Beijing protects its sea lanes for strategically vital materials by fortifying certain islands in that broad expanse of ocean. It's also interested in the vast petroleum reserves that lie beneath them.

While America reshuffles priorities every four years, the PRC has doggedly stuck with the same long-term plan. The *Economist Magazine* offers the most revealing look at the veiled aggression. Because of Communist China's expanding economic interests, it considers this huge body of water a trade zone so exclusive as to be guarded by the PLA (People's Liberation Army) Navy.[2] Seven sovereign nations have suffered from it: Japan,[3] Taiwan, the Philippines, Vietnam, Malaysia, Indonesia,[4] and Brunei.[5]

While energy exploration in the Spratly Islands of the South China Sea has so far produced very little, they are still thought to be sitting atop huge reserves of oil and natural gas.[6] So, China wants this string of islets, reefs, and shoals for more than just shipping-lane protection. (Look again at Map 1.1.) Of the 100 Spratlys, the PRC had occupied seven by 1999, Vietnam retained physical control over 27, the Philippines eight, Malaysia seven, and Taiwan one.[7] In 1988, China sank three of Vietnam's ships and drove its troops off Fiery Cross Reef. Four years later, the National People's Congress of China claimed sovereignty over all Spratlys and authorized the PLA Navy to "evict trespassers."[8] In 1995, Beijing took over Philippine-owned Mischief Reef.[9] It has since dredged that reef"s entrance to create a harbor for large ships. It has also fortified that harbor against foreign assault.[10]

But, the PRC's designs on this region go back farther than that. In 1974, Chinese troops forcefully evicted South Vietnamese forces from some of the South China Sea's Paracel Islands.[11] On Woody Island, it has since built port facilities, ammunition bunkers, fuel tanks, gun emplacements, and a runway long enough for its largest aircraft.[12] Taiwan also lays claim to some Paracels.

U.S. military intelligence personnel claim that what has happened to the Spratlys and Paracels is part of "China's 'outward expansion' to move forward on its 'island chain strategy' . . . to control all the territory stretching in the huge arc from Japan to Indonesia."[13]

American Efforts to Stem the Tide

While Washington seems only mildly perturbed by China's warlike activity in this vast swath of open ocean, it continues to insist upon unobstructed travel in the region for all nations. As recently as February 2021, it dispatched naval contingents its most heav-

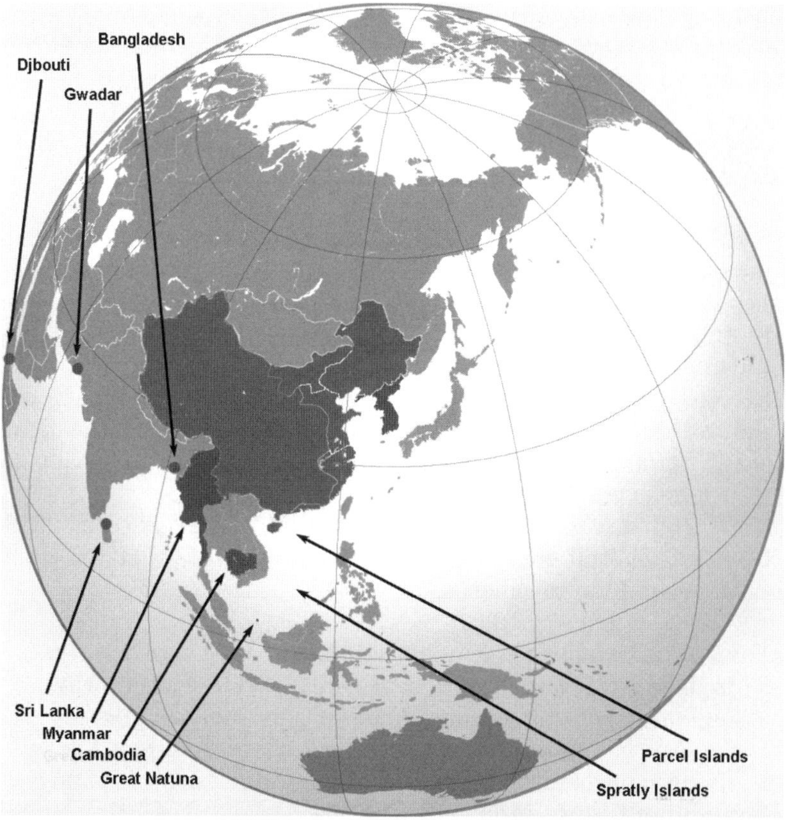

Map 1.1: PRC Allies and Bases That Protect Main Shipping Lane
(Source: "File:Greater Asian Co-Prosperity Sphere.png" at WIKIMEDIA COMMONS, created by Kendrikdirksen, all Japanese occupied areas replace by Chinese,"©)

ily defended parts.[14] In October of 2015, a U.S. destroyer sailed to within 12 nautical miles of "one of China's man-made islands" in the Spratlys. Most prominent are those at Subi Reef and Mischief Reef. At the latter, China is now building runways for military aircraft.[15]

In the Parcels, the PRC has had advanced fighter jets stationed at a fully operational airfield on Woody Island since November 2015.[16] In February 2016, *NPR News* reported HQ-9 surface-to-air missiles on nearby Tree Island. Such armaments are capable of shooting down any airplane within 100 miles. Even more disturbing was the *NPR* announcer's final comment—this was the PRC's "sovereign territory" and America "will have to get used to it."[17]

How the PRC Defends Its Actions

China has been calling its land reclamation projects in the Spratlys a "public service." But, her neighbors perceive them as something else—airfields from which the PRC may enforce an illegal claim over the entire South China Sea.[18] One might expect a few dump truck loads of rock at the end of one or two prospective runways. But, this construction project has produced some "3,000 acres of artificial islands in the South China Sea using sand dredged from the ocean floor."[19] That's enough landmass for many long airstrips, each with its own harbor and ring of defensive emplacements.

Though dredging took place near Fiery Cross Reef in May 2016, the PRC's overall effort in the Spratlys has also produced many above-ground improvements.

> The Defense Department said three of the land features in the Spratly Islands now have nearly 10,000-foot runways and large ports in various stages of construction.
>
> And it [China] has excavated deep channels, created and dredged harbors, and constructed communications, logistics and intelligence gathering facilities.[20]
>
> — *MSN News* (from AP), 14 May 2016

There Has Been a Surprisingly Rapid Transformation

In May 2016, *Time Magazine* published a detailed satellite photograph of how much Fiery Cross Reef had changed in two years. What had been a solid reef with a tiny weather station on two exposed rocks was now a deep water port surrounded by a full-length military airstrip and every support facility imaginable. In effect, the Chinese had dredged out the center of the reef to create a more solid periphery.[21] According to *Time,* this is just one of seven artificial islands built by the Chinese since 2014. After radar and surface-to-air missile batteries were added to the defense of Woody Island in the Paracels, the Head of the U.S. Pacific Command accused the

PRC of "militarizing the South China Sea." If what had previously been Fiery Cross Reef were further reinforced with Dongfeng 21D anti-ship missiles,[22] it would severely threaten Free World shipping. Concentric rings of similar bastions between Hainan Island and the Straits of Malacca might guarantee an uninterrupted flow of petroleum and minerals from the Middle East and Africa back o China in time of war.

After revealing a Chinese commentators' on-line plans for turning formerly Philippine-occupied Scarborough Shoal into another "weaponized" artificial island, *Time* disclosed the PRC's method of takeover. First, scores of Chinese fishing boats show up with a couple of "coast guard" escort ships. Then, "fishing militiamen" storm ashore. Among the trawlers at the Tanmen port on Hainan Island are some big "military boats." When President Xi visited Tanmen in 2013, he first "praised the local maritime militia, composed of well-subsidized fishermen and decommissioned soldiers, for plying distant waters." Then, according to State media, he urged its members to "support the construction of islands and reefs." Of course, this was the same militia that had seized Scarborough Shoal in 2012 after a failed diplomatic effort by Washington to support the Philippine government.[23]

A World Court Makes Its Ruling

In July 2016, the Permanent Court of Arbitration in the Hague made its decision on a case lodged by the Philippines in 2013. It ruled that China's historic claims to the South China Sea "had no legal basis." Because Communist China could no longer claim any territorial rights to the reefs and shoals converted to artificial islands, its leaders ignored the panel's decision.[24]

The PRC was then to test Western resolve by fully developing an artificial island at Scarborough Shoal (also claimed by the Philippines). Occupied by the PLA since 2012, this shoal had already received a Chinese name—Huangyan Island.[25] Though that July 2016 Hague tribunal had specifically "ruled that the Scarborough Shoal's surrounding waters were a common fishing ground and censured Chinese-imposed restrictions on Philippine fishermen straddling the area,"[26] Chinese ships were to mass there in early September 2016.[27] According to a distinguished member of the Philippines' Supreme Court, they were doing more that fishing.

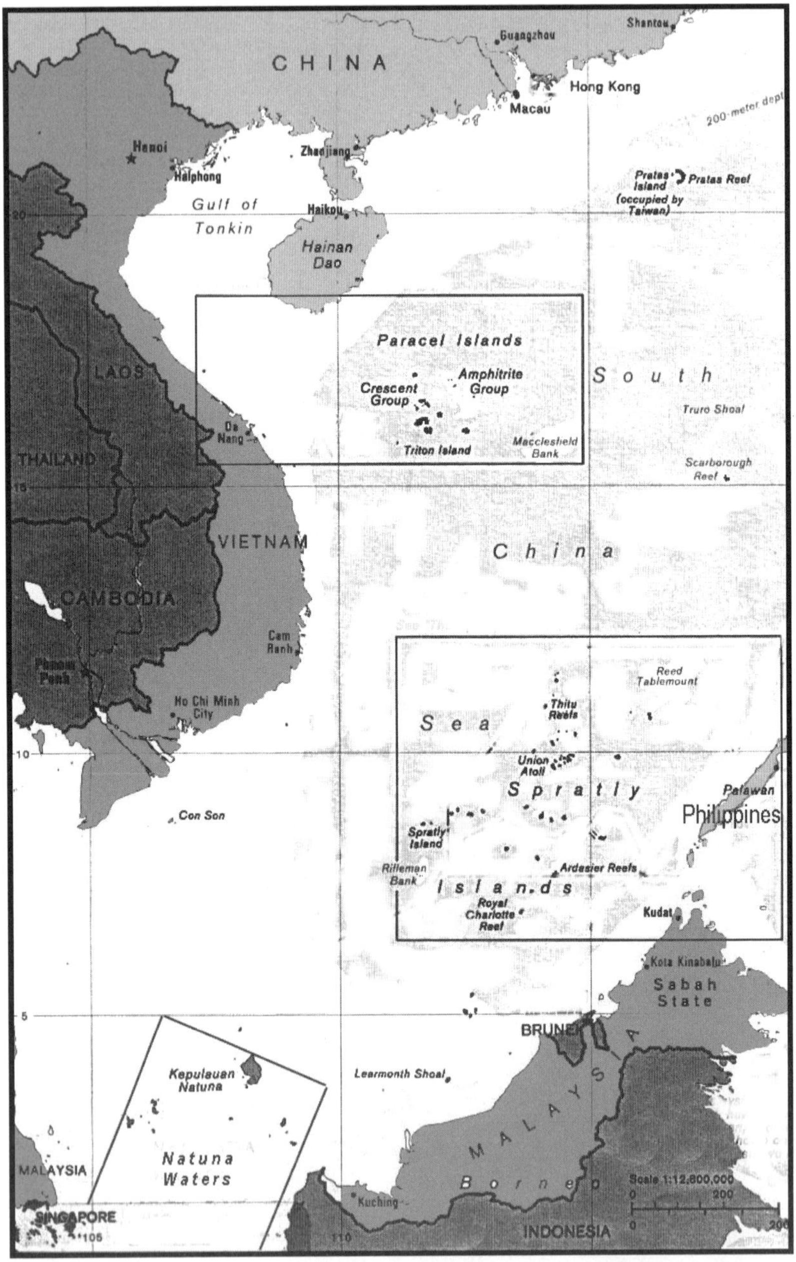

Map 1.2: China's "String of Pearls" in the South China Sea

(Source: Courtesy of General Libraries, Univ. of Texas at Austin, from their website for map designator "schina_sea_88.jpg")

> Retired Senior Associate Justice Antonio Carpio on Tuesday warned China may "very soon put up" air and naval bases on Scarborough Shoal. . . .
>
> Carpio said that without an air and naval base, ADIZ [Air Defense Identification Zone] cannot be enforced over the South China Sea because of a "hole in China's radar, missile, and jet fighter coverage" in the vicinity of Scarborough Shoal.[28]
>
> — *CNN Philippines,* 9 June 2020

Likely Target at South End of the PRC's Personal Ocean

For full access to the Straits of Malacca, the PRC would need allies or outposts all along the coastlines of Malaysia and Indonesian Sumatra. It would also occupy a few islands at the Straits' eastern entrance. At the very edge of the South China Sea, the Anambas and Natuna Besar also have access to its petroleum reserves.

Indonesia's Natuna Island Regency

The Anambas and Natuna Besar (a.k.a. Great Natuna) both have some elevation to them. Great Natuna is mostly composed of easy to dig sandstone and limestone. Its highest point is Mount Ranai at 3,396 feet. Imperial Japan occupied Great Natuna as part of its Greater East Asia Co-Prosperity Sphere in World War II (WWII).[29] That means the island is still laced with underground emplacements and passageways. While now outposted by 1000 Indonesian soldiers, it could be quickly seized by the Chinese PLA in time of war. At that point, the U.S. Marines might be called upon to take it back. To do so, they would need every bit of what's been learned from previous wars in the region.

Great Natuna is at the same latitude, but far to the west of Peleliu. Both are comprised of coral limestone forced upwards by sea-bottom pressures. Through rainwater run off, such rock develops natural caves and crevices—so many that Peleliu became nearly impossible to seize in WWII. Now, other PLA bases along either side of the Malacca Straits may have enough protruding coral to allow for a Peleliu-like defense. (Look back at Map 1.1.)

China's "String of Pearls" Strategy

Most Chinese bastions within the South China Sea are on tiny islets that barely rise above the surface of the ocean. In time of war, their aircraft-launching capabilities could be easily controlled by America. But, with enough cruise missiles hidden below ground and then fired remotely, they would still pose quite a threat to any U.S. warship within hundreds of miles.

> Beijing has fortified multiple forward operating bases [in the South China Sea] with anti-cruise missiles (ACMS) and surface-to-air missiles (SAMS) along with underground storage facilities, hangars, radars, and sensor arrays.
>
> PLAN [PLA Navy] modernization also encompasses the deployment of . . . land-attack cruise missiles (LACMs).[30]
>
> — Rand Corporation study, December 2019

The latest U.S. bunker-busting munition can only stop what has been precisely located. There would be no heat signatures to help find these island-embedded missiles. The only sure way to remove them would be an invasion. That might not be as difficult as at Peleliu, but it could easily rival the Tarawa landing.

The U.S. military now bases much of its warfighting ability on the latest technologies for target acquisition and destruction. It no longer practices the tactical techniques for seizing a heavily fortified island from a skilled defender. That makes its power to do so a dangerous illusion. This book discusses the extent of the shortfall and how to fix it.

2 The Bairoko Incident

- What made Japanese defense of Bairoko so daunting?
- Why were Marine Raiders happy after failing to breach it?

4th Raiders had learned how to win with maneuver alone.

(Source: Govt. recruiting poster, image designator "post_usmc_168th-birthday_ww2.jpg," retrieved from U.S. Nat. Archives and Records Admin., by www.bluejacket.com)

Iwo Jima's Defense Line Prototype at Bairoko

At 0100 hours on 18 July 1943, all 700 men of the Maoist-oriented 4th Raider Battalion debarked their destroyer-transports at Enogai Harbor on the western end of New Georgia. As part of 1st Marine Raider Regiment, they were to attack Bairoko Harbor on the other side of Dragons Peninsula one day later. It was through the barge base at Bairoko that the Nipponese had been reinforcing their airfield at Munda.[1]

Foreseeing a ground assault from the east, Bairoko's garrison had constructed a new type of defense in depth. It was not only to

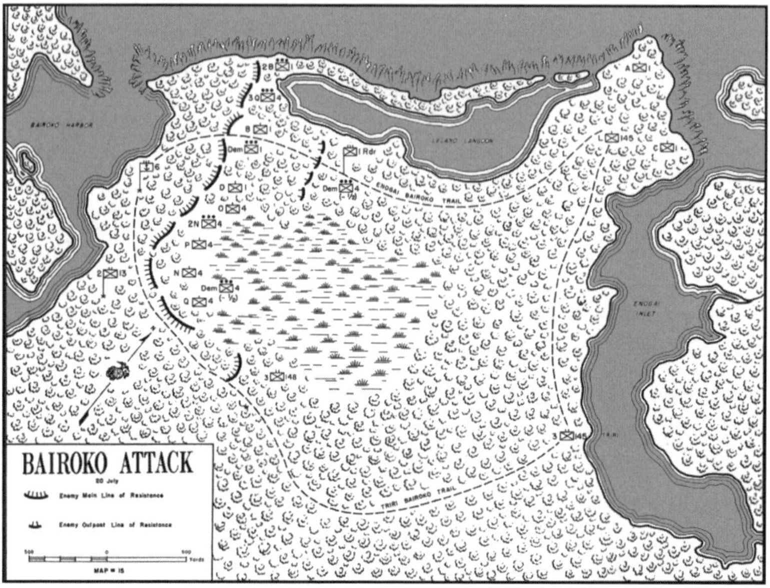

Map 2.1: Defenses Perfectly Meshed with Terrain at Bairoko
(Source: "Marines in the Central Solomons," by Maj. John N. Rentz, USMC Historical Monograph, HQMC, 1952, map 13.)

stymie the Marine Raiders' attack, but also the U.S. War Department's appreciation for light-infantry tactics in general. (See Map 2.1.)

After a difficult approach march and "no-show" airstrike, the Leatherneck commandos finally made contact with the Bairoko defenders about 1000 hours on 19 July.[2] For the rest of the day, they fought—with 1st Raider Battalion on the right and 4th Raider Battalion eventually on the left. Blocking their path were four back-to-back belts of mutually supporting machinegun bunkers.[3] (See Figure 2.1.)

After an attached U.S. Army battalion's attack from the south failed, the Raiders were forced to retire without ever penetrating all four belts. The reason later proffered by the Joint Chiefs of Staff (JCS) was that "lightly armed troops cannot be expected to attack fixed positions protected by heavy automatic weapons, mortars, and heavy artillery."[4] Such an opinion was fully consistent the Industrialized World myth that light infantrymen are unable to "hold their own" in conventional combat.

Tactical Evolution Since WWI

In July of 1943, most deployed units of the U.S. Marine Corps (USMC) were still conducting the traditional 2nd Generation of Warfare (2GW). They went after the enemy's manpower concentrations with as much firepower as possible. The 3rd Generation of Warfare (3GW) entails bypassing the foe's frontline defenders to more easily get at his command and logistics networks.[5] Among the few U.S. outfits dabbling in 3GW were Carlson's 2nd Raider Battalion and Roosevelt's 4th. While trying to emulate Mao's Eighth Route Army, they had replaced the strict headquarters control

Figure 2.1: Coconut-Log Pillbox
(Source: TM-E 30-480 [1944], p. 161.)

of 2GW with more bottom-echelon initiative. On the famous "Long Patrol" at Guadalcanal, Carlson allowed tiny contingents to choose their own schemes of maneuver during swarm-like offensives.[6] And the 4th Raider Battalion at Bairoko shared the same mindset. The Regiment's "top-down" plan of attack had fallen through when the artillery and air support failed to materialize. But still available to the 4th was a "bottom-up" alternative. It involved fireteam opportunism and a *gung ho* (working together) means of control.

This New Way of Fighting Fared at Bairoko

On that fateful day on new Georgia, the more conventional 1st Raider Battalion made first contact with the enemy strongpoint matrix. While light-infantry oriented, that Marine unit had yet to adopt any of the advanced assault techniques of the Eastern World—like the short-range infiltration of German Stormtroopers or Asian sappers. All 1st Raider units simply came on line with the initial row of bunkers to maximize their firepower. Then, one unit attacked a single bunker while the rest provided covering fire. Once that belt had been breached, the same assault unit tried to work sideways along it. But, a second belt protected the first and made such a flanking maneuver nearly impossible.

Only then would the 4th Raider Battalion get its chance at the objective. Through trial and error, it finally came up with a better way of doing things. Its final desperate attack produced a tactical innovation that would help to win the crucial battle of Okinawa.[7]

Challenges Facing the 4th Raider Battalion

The 4th Raiders would attack their part of the strongpoint array all afternoon. When darkness finally forced them to withdraw without fully piercing that array, its members became quite proud of their "failed mission." What did those frontline Gyrenes realize that their rear-area leaders had yet to understand? Fortunately, there are two detailed accounts of the action.

The Bairoko defenses had never been reconnoitered,[8] nor had any Raider adequately rehearsed assaults against mutually supporting positions. The standard bunker take-down for this period in USMC history was to "blind 'em, burn 'em, and blast 'em." But the 4th Raiders had no flame throwers, so they could only crawl forward to toss in explosives.

In other words, those 4th Raiders came very close to hastily attacking a "prepared enemy position"— a no-no in the world of tactics. This had been no haphazard defense, but a prototype of what would later create so much havoc on Peleliu, Iwo Jima, and Okinawa. It consisted of four slightly offset strongpoint belts on parallel coral outcroppings. Each pillbox had its own heavy machinegun, riflemen outguards, protective fire from neighboring pillboxes, and "coconut-log-and-coral" construction.[9] (Look back at Figure 2.1.) With snipers in trees and "light *Nambu* machineguns in outpost foxholes,"[10] the bunker complex also enjoyed an early warning system that obscured its precise location. Because Japanese emplacements had been placed on either side of each coral berm,[11] there may have been tunnels or a covered trench between them.[12] Such passageways would have permitted a fallback option.

What Had Happened from an Evolutionary Standpoint

Highly skilled and then loosely controlled light infantrymen should be able to assault a fully supported deliberate defense without any tank, artillery, or aircraft support. And that's what the 4th Raiders ended up doing at Bairoko. By all indications, some of their units briefly penetrated three of the four defensive belts through little more than fireteam cooperation.[13] With a few nighttime assault techniques, they could have finished the assault on Bairoko Harbor.

But, such ideas have been difficult to entertain in the Free World's burgeoning arsenal. Only after poorly armed militiamen did precisely that in the late 1960's would anyone take such talk seriously.

The Initial Contact with the Japanese Defense Line

As the U.S. Raiders approached the Bairoko bunker system, they were sniped at from treetops with light machineguns.[14] Then, two Marine companies pushed on line through "a series of outguards" to reach the main defenses.[15] That means the Japanese had deployed

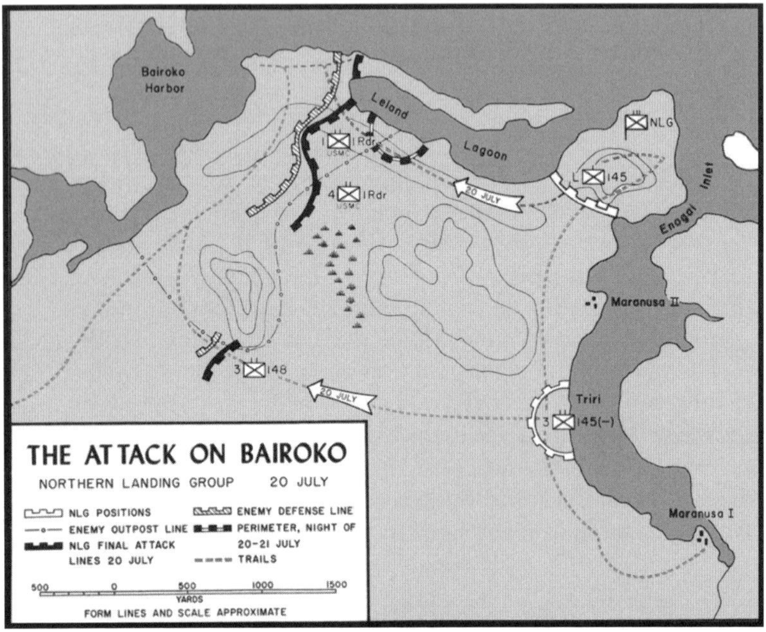

Map 2.2: 4th Raider Battalion Penetrates Bunker Belts at Bairoko
(Source: "From Makin to Bouganville: Marine Raiders in the Pacific War," by Maj. Jon T. Hoffman, Marines in WWII Commemorative Series, HQMC, 1995, p. 34)

special skirmishers to the front of a Main Line of Resistance (MLR). Those skirmishers functioned as roving sentry posts and tiny ambushes. This was not the first time such a sophisticated early warning system had been tried. During the Nomonhan Incident of 1939, Japanese units used defensive scouts talented enough to "shadow" enemy infiltrators.[16]

But, this unexpected machinegun fire was a warning of things to come on subsequent islands, for ambush was the core principle of a 3GW-oriented defense.

Specifics of the Penetration Attempt

Upon arrival at the objective, those Raider companies had other problems. Not only was each roofed emplacement protected by the machinegun fire of those behind it, but it had also been preregistered by Japanese mortars.

The 1st Battalion attacked the enemy fortification for a while, then Regimental commander committed the 4th to its left-hand portion. He had wanted to outflank all or part of the strongpoint array, but that was not to be. However, it may have prompted the unconventional 4th to partially envelop each bunker.

> Liversedge, accompanying the northern prong of his offensive, committed the 4th Battalion in an attempt to turn the enemy flank, but it met the same heavy resistance [as the 1st Battalion.] The [Maoist] raider companies slowly worked their way forward, and by late after noon they had seized the first two enemy lines. However, throughout this advance enemy 90mm mortar fire swept the Marine units and inflicted numerous casualties.[17]
>
> — *From Makin to Bougainville,* by Hoffman
> Marine Corps History Center, 1995

What Roosevelt's Raiders eventually achieved to become so elated will take a detailed inspection of the battle chronicles. It likely has to do with a smaller maneuver element.

> Approach to Bairoko was to be made by two columns. Two full-strength companies (B and D) of the 1st Raider Battalion and the four companies of the 4th Raider Battalion were to make the main effort, advancing along the south shore of Leland Lagoon straight toward Bairoko and the north flank of the Japanese positions. Schultz' [Army] battalion was to move from Triri toward Bairoko to hit the south flank of the Japanese positions. . . . The departure time was set for 0730 [on 20 July], with an airstrike scheduled for 0900 to precede the actual attack on the harbor. . . .
>
> Confronting the Raiders was a series of log and coral bunkers dug into the rising ground under banyan roots, and well camouflaged with palm fronds and branches. The ridge ahead blazed with [enemy] fire. . . . [T]he emplacements supported each other with lanes of interlocking fire. Further protection was furnished by Japanese soldiers in trees overhead who sniped at the Marines with Nambu (0.25 caliber) light machine guns. . . .
>
> By noon, the first line of enemy resistance crumbled,

broken under the relentless pressure of the [1st Raider Battalion's] . . . units. Unable to use the 60mm mortars because of the jungle canopy, lacking the new flame-thrower weapons, and without air or artillery support, the Marines breached Okumura's defensive line by knocking out first one pillbox and then another by demolitions and overwhelming small-arms fire. But losses were heavy and progress was slow. . . .

Shortly after noon, , with the 1st Battalion clearly needing quick assistance, Liversedge committed Currin's [4th Raider] battalion to the fight. Company P (Captain Walker) was in close support behind Griffith's [1st] battalion, and thus able to move quickly into the line. Kemp's Company D . . . was receiving heavy fire on its left flank and Walker now attacked toward this opposition. . . .

Walker's fresh company . . . was barely able to move forward before criss-crossing fire from both right and left flanks held it back. While Walker scouted his front lines to determine the location of the machine guns facing him, Captain Snell moved his Company N into position behind Walker's unit to refuse the left flank and support Walker's attack. . . .

In the next two hours, the Raider attack slowly punched through *two different defensive lines*, uncovering a number of *bunkers on the reverse slopes*. Company D, riddled with casualties by the heavy and continuous enemy fire, scrambled to the top of a ridge line which overlooked the harbor at Bairoko, about 500 yards away. But between the Raiders and their objective lay another series of formidable fortifications. . . .

At 1445, sporadic but accurate *mortar* fire from enemy positions on the inlet suddenly changed into an intense barrage that shook the attacking lines. The Marines, without weapons for counterbattery fire, could only press closer into their shallow positions behind scant cover on the ridge lines and try to weather the pounding. Estimated to be 90mm rounds, the shells inflicted further casualties, mainly from tree bursts overhead. The barrage was immediately followed by a screaming counterattack. . . .

Unfortunately, there were no ready reserve units. Nothing had been heard from the Army battalion which was supposed to hit the south flank of the enemy, but sounds of firing from that direction indicated that Schultz [its commander] was engaged. . . .

Following Company D's return to its former position, the 4th Battalion found movement easier, and Companies N and P managed to move forward in the face of stiffening fire to extend the NLG [Northern Landing Group] lines more to the southwest. But the move was costly; both companies received heavy casualties. Company Q (Captain Lincoln N. Holdzkom), the sole remaining company as yet uncommitted, moved up to the rear of the other three 4th Battalion units to be in position for an attack when directed.

By 1600, the Japanese had been pushed, still defiant and dangerous, into an area on the Bairoko Harbor headlands about 300 yards wide and 800 yards long. Their back to the sea, the enemy defenders kept up a sustained and murderous machine gun and mortar fire that showed few signs of slackening. In an effort to strike one last, conclusive blow, Liversedge ordered Company Q into the lines. Holdzkom's company moved around the left flank of Company N in an attack straight into the teeth of heavy enemy fire. . . .

The outcome of Liversedge's last attempt to take his objective was not long in doubt. Despite the vigor of Company Q's attack, the overwhelming fire of the enemy won. Badly depleted in a matter of moments, Company Q was forced to retire. Repulsed, the company reeled back, virtually noneffective through its losses. The tactical situation had been opportune for one last heavy punch to knock out the enemy defenders, but without artillery, air support, or other heavy weapons, the Raider battalions could not deliver it.

For Liversedge, Schultz' failure to attack aggressively on the left flank was the final blow in a series of sharp disappointments. . . .

By this time the Raiders (1st and 4th) had nearly 250 casualties, or about 30 percent of the force. We had another 150 men tied up getting them evacuated to aid stations and to Enogai. There was nothing to do but pull back.[18]

– *Isolation of Rabaul,* by Shaw and Kane
Historical Branch, HQMC, 1963

Fireteams Were at the Heart of the Innovation

By 1400 hours, the two Raider battalions had punched their way through two of the four defense lines. No U.S. supporting arms had been utilized, nor would they be for the rest of the day. In an attempt to create a tiny breach in each enemy belt and then move sideways along it, the light-infantry-oriented 1st Raiders had attacked one bunker at a time with a platoon or squad. The rest of the battalion suppressed all other bunkers with small-arms fire.

It did not take the GW-trained 4th Battalion long to realize that no part of the defense matrix could be enveloped—neither its left end nor any internal bunker grouping. In other words, there could be no moving sideways from a narrow beachhead. In a final desperate thrust, Company Q attacked in the late afternoon. It likely went after all visible bunkers at the same time—giving each fireteam its own parallel lane. In that way every bunker in the subsequent rows could be engaged by either suppressive fire or frontal assault. That "more decentralized" way grinding forward would eventually run out of daylight, but its specifics were what the 4th Raiders would become so enamored. They had discovered a safer way to overcome heavy enemy resistance.

According to the chronicle, the 4th Raiders advanced very slowly. That allowed for some forward investigation (the precursor to "recon pull"). It also permitted the "visualization" of assaults for which there had been no rehearsal. All fireteams then witnessing a successful few bunker assault would have a pre-tested method. They could now conduct a "quick deliberate attack"—just like the German Stormtrooper squads had in the Spring Offensive of 1918.[19] Through less overall control, collective cooperation, impromptu reconnoitering, and on-the-job practice, those U.S. fireteams could do what their headquarters had deemed impossible. They could penetrate a heavily fortified enemy MLR without any risk from friendly (but error prone) tanks, artillery, or aircraft.

The bottom-up-focused 4th Raiders already knew the dangers of going directly at a bunker. So, without straying from each fireteam lane, one buddy team would have crawled left through the microterrain while the other went right. In that way, the separate pairs of riflemen could have played off each other's movements—advancing only after their quarry was distracted. They were also in a better position to lend base-of-fire assistance to fireteams in adjacent lanes.

This tiny-element maneuver must now be remembered by all U.S. units attempting to beach a modern defensive belt. The 4th Raider Battalion's direct descendent would likely use it to capture the front face of Crescent Hill in Okinawa's Sugar Loaf Complex.[20] Without it, the infamous Shuri Line might never have fallen. That makes this *"gung-ho*-like" Marine stratagem a viable alternative to the dangerous steamroller approach. Overwhelming firepower would not be enough to breach North Korea's new border defenses in 1953.[21]

Late in the day at Bairoko, the foe's "trump card" was played. Highly accurate mortar fire came raining down on the bunkers after all exterior outguards had taken cover inside. That intruder-discouraging shrapnel would have helped bunker occupants to pull back through a trench or tunnel. But, the 4th Raiders ran out of daylight before any such move was discovered.

The Importance of This Precedent Cannot Be Overstressed

The 4th Raider Battalion would help to "mop up" Enogai Inlet for a month before heading back to Guadalcanal. Despite the 156 casualties at Bairoko,[22] its morale remained high. Its men had demonstrated how a little GW training had permitted Maneuver Warfare (MW, a.k.a. 3GW) at the fireteam level—something modern-day Marines still aren't allowed to do.

The battle's official account confirms the "recon pull" part of the method: "*Walker* [of Company P] scouted his front lines to determine the location of the machine guns facing him."[23] No one disputes the impromptu development of a viable tactical technique. So, a series of quick deliberate attacks had, in fact, occurred. But, unlike the German Stormtrooopers assault method of WWI, there had been no supporting-arms fire necessary.

Without airstrikes, artillery, heavy mortars, flame throwers, tanks, or pre-arranged assault technique, those Mao-oriented U.S. Raiders had gotten three-fourths of the way through what is still the "state of the art" for a rural defense.

With the self-confidence that comes from guerrilla training and cooperation born out of *gung ho* spirit, the Raiders had done the impossible. Not until an Asian army with a few amphibious tanks, limited artillery, and no aircraft persevered against America's full might in Vietnam would there be a better example of what properly prepared grunts can do. U.S. troops are too prideful to copy a foe, but they may be willing to emulate a Carlson or Roosevelt Raider.

This Had Not Been an Isolated Incident

While attempting to seize Munda Airfield in July of 1943, U.S. Army personnel reported bunker systems like those at Bairoko. Those Munda emplacements were connected by communication trenches.

> A line of pillboxes stretching from Laiana beach northwest for more than 400 yards had been breached. Typical of the defenses was a cluster of seven pillboxes which covered a frontage of only 150 yards, each position defending and supporting the next. Overhead and frontal protection consisted of two thicknesses of coconut logs and three feet of coral. Skillfully camouflaged, with narrow firing slits, the bunkers were virtually a part of the terrain and surrounding jungle. . . .
>
> . . . [E]nemy soldiers were dug in and covered by low, two-level camouflaged coral and log emplacements with deadly interlocking fields of fire. Trenches bulwarked by coconut logs connected the bunkers.[24]
>
> – *From Makin to Bougainville,* by Hoffman
> Marine Corps History Center, 1995

In August of 1943, the Bairoko Harbor fortifications were finally captured by the U.S. Army. The entire garrison had been evacuated by ship.[25] Other Japanese-occupied islands farther north would present similar challenges: (1) mutually supporting bunkers covered by indirect fire; (2) special skirmishers out in front of the bunker belts; and (3) the unexpected withdrawal of troops supposed to fight to the death.

3 The Assault on Biak

- Were U.S. Army planners aware of all invasion problems?
- What would the unopposed landing force later face?

41st Infantry Division "Jungleers" on Biak.

(Source: "Jungleers on Biak," by K.Rocco, and from this url: https://www.nationalguard.mil/Resources/Image-Gallery/Historical-Paintings/Heritage-Series/Jungleers-on-Biak/)

The Battle for Biak

U.S. Army units took from 27 May to 7 September 1944 to seize this overheated tropical isle off the northwestern coast of New Guinea. As part of Gen. Douglas MacArthur's WWII island-hopping campaign, it would help to clear a path to the Philippines. But, stratagems used by the IJA (Imperial Japanese Army) to protect Biak will be forever remembered. This had been the Yanks' first full exposure to Japan's new defense plan.

There are two ways of viewing this battle: (1) from the "big-picture" standpoint; and (2) from a close-combat perspective. To

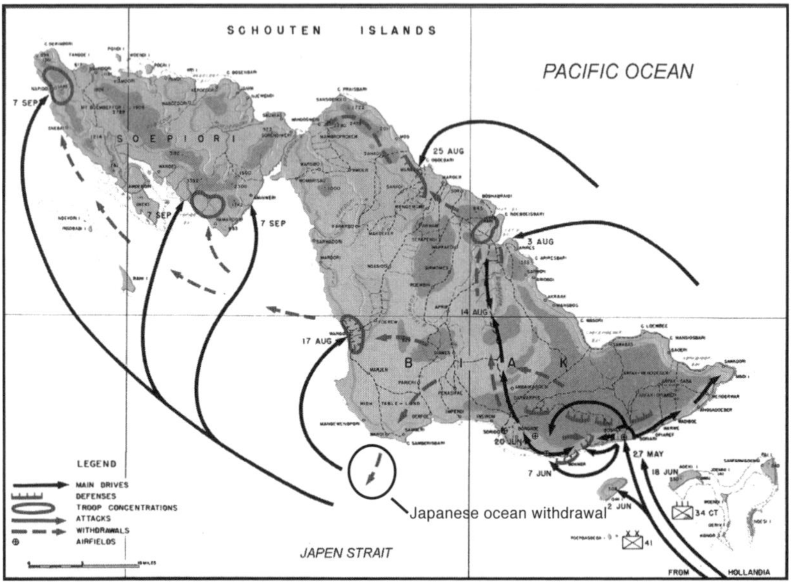

Map 3.1: MacArthur Reported an Enemy Withdrawal from Biak
(Source: "Biak Island Operation ...," The Campaigns of MacArthur in the Pacific, vol.1, U.S. Army Ctr. of Mil, Hist., Wikimedia Commons, public domain.)

learn anything new, the latter approach must be taken. Something unusual had happened at Biak. Map 3.1 from Gen. MacArthur's official files shows Japanese soldiers shifting rearward—some via boat—until 7 September. It also shows some airfield defenders moving back out into the open ocean. Those are things the die-hard Nipponese were never supposed to do.

Biak's Geography

This medium-sized island was largely covered with equatorial rain forest. It also had some elevation to it. At its southeastern end, low coastal ground quickly rose into coral limestone terraces and finally an inland plateau. "Within those terraces and cliffs lay countless caves linked by fissures and tunnels."[1]

At Biak's southern coast was a flat coral shelf with very little topsoil.[2] That made it hard for U.S. infantrymen to dig in for the night. (The Marines about to invade Peleliu could have used more information about this aspect of Pacific islands.) Biak also had an unusual watershed system that first absorbed most rainwater into porous ground. Sub-surface tributaries then made drinking water almost impossible to find.

> The south-central and eastern areas of Biak are without surface streams, as the terrain is inclined to subterranean streams and pools. Springs and water points are rare and their possession was of high importance throughout the operation. The Japanese defense of Biak was based on brilliant appreciation and use of the terrain.[3]
>
> — "History of the Biak Operation," 41st Division

> Little fresh water is readily available on Biak, since most of the streams run through underground channels that drain even the heaviest rainfall from the surface.[4]
>
> — "Approach to the Philippines," by Smith
> U.S. Army Center of Military History, 1996

Thirsty troops are less productive in hot climates. But, the invaders of Biak had a much bigger problem with regard to the terrain. Just beyond Mokmer Airfield, rainwater runoff had created a labyrinth of sub-surface caverns and connecting galleries. This subterranean network was most easily reached through three big sinkholes. (See Figure 3.1) When ringed with surface emplacements, these natural entrances to Biak's underground city provided its garrison with tactical options the Bairoko defenders hadn't enjoyed. One was to shell Mokmer Airfield without any muzzle flashes. Another was to secretly enter any approaching U.S. formation. Japanese marauders could go any place a metropolis sewer system would have taken them. That's why—at Biak—urban warfare discoveries from Europe may have helped to shape Japan's new island defense plan. It was soon to get more productivity out of limited assets.

How Far City Fighting Had Progressed by 1944

At Stalingrad in early 1943, German units assaulting holdout enclaves along the Volga River had been drawn into kill zones, to

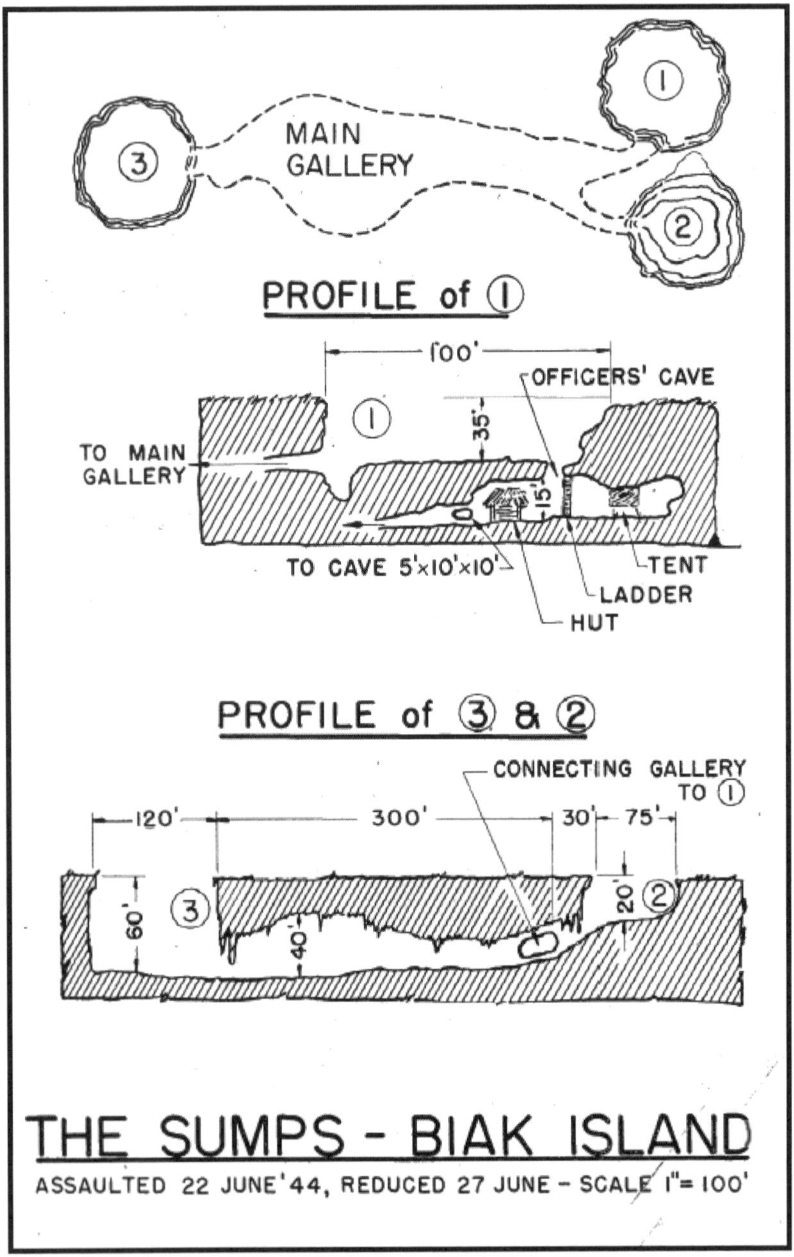

Figure 3.1: At the Center of the "West Caves"

(Source: "History of the Biak Operation 15-27 June 1944," U.S. 41st Army, 1944, Armed Forces Staff College Library copy, p. 11.)

which semi-autonomous Russian squads could swarm. Other Nazi soldiers became canalized while moving through rows of fortified buildings. They soon found themselves inside the firesack of a "soft" defense matrix.

Such traps came to define a new way of contesting attack that came to be known as "retreat combat." It would be practiced by the Germans at Berlin and Japanese at Manila.[5] There, subway and sewer ducts enhanced the withdrawal process. Its rural equivalent isn't hard to imagine. With a little runoff channel enhancement, a coralline-limestone landscape would permit the shift of unused manpower and ordnance rearward.

What Biak Further Revealed About the Foe's New Strategy

At Bairoko, the Japanese had resisted through mutually supporting rows of bunkers. U.S. scouts had seen trenches being dug in the area, but no bunker occupants falling back to a subsequent row. Yet, forward-deployed Japanese riflemen in trees and light machinegunners in outpost foxholes did "retire" to the first row of bunkers.[6]

Nor at Biak would the rearward movement of emplacement occupants be witnessed, though trenches between rows existed. Only the retrograde movement between cave complexes can be verified. But, the Biak opposition would clearly perform the other aspect of retreat combat—drawing U.S. forces into preplanned kill zones. How they did so was Asian in origin. (See Map 3.2.)

> Due to the size of the coastline he had to defend, and based on an appreciation of the Allied objectives, Kuzume [the Japanese commander] focused his defensive plans away from the water's edge. Instead, he decided to carry out a feint, allowing the Americans to come ashore unopposed so that they would advance into the trap he had prepared for them utilizing a series of caves that were located [just] west of Mokmer and to the east of Bosnek. This defensive complex was intended to turn the area around the vital airfield into a honeycomb of defended caves and pillboxes filled with riflemen, automatic weapons, artillery, batteries of mortars, and a single company of Type 95 Ha-Go light tanks. The western caves were connected by a series of underground

tunnels and were largely constructed for fighting purposes. Kuzume also stockpiled these positions with ammunition and other supplies, with several dumps located around the eastern cave area, along with living quarters for the defenders (R.R. Smith, *Approach to the Philippines,* U.S. Army Ctr. of Mil. Hist., 1996).[7]

— *Wikipedia Encyclopedia,* s.v. "Battle of Biak"

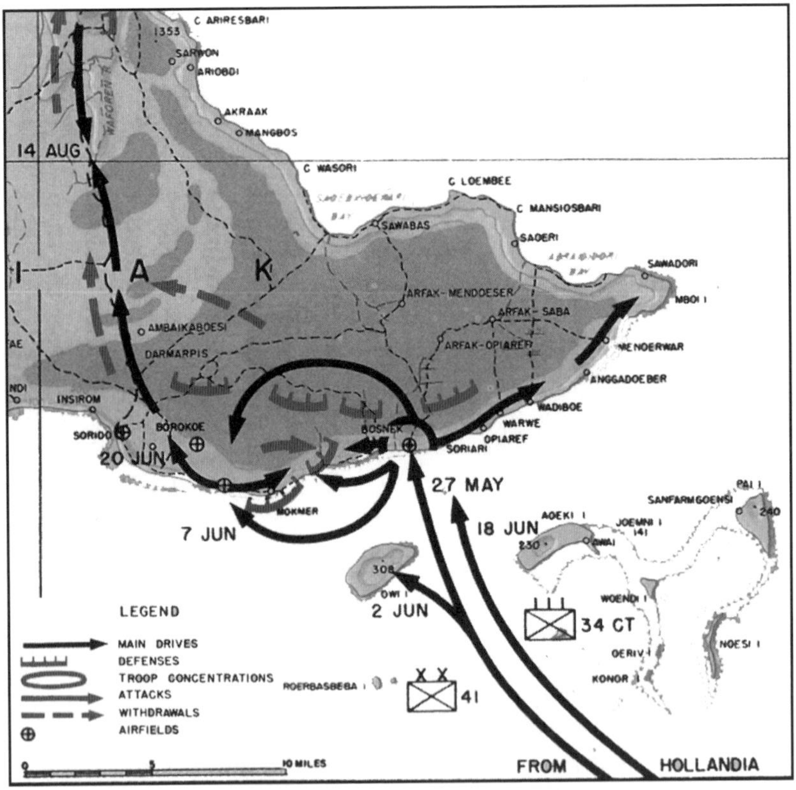

Map 3.2: The Mokmer Trap

(Source: "Biak Island Operation ...," The Campaigns of MacArthur in the Pacific, vol.1, U.S. Army Ctr. of Mil. Hist., Wikimedia Commons, public domain)

The Soon Famous Ambush

There's an ancient Chinese snare that wouldn't become well known until the Korean and Vietnamese Wars. It has been variously called the "Closing V" or *Haichi Shiki* battle array. The initial struggle for Biak's airfields may provide its first recorded example during WWII. There, a string of cave complexes had formed the top leg of a leaning "V," with the shoreline itself comprising the opposite leg. Mokmer Aerodrome was just beyond the point of this V and could thus function as bait for any intruder to enter it. (See Figure 3.2.)

In Japanese, *haichi shiki* means "bowl formula." Something like it must have been applied to the Mokmer drive, because that drive's forward participants had to be rescued by landing craft. Originally for countering a pursuer, the *Haichi Shiki* ambush would be more often noticed in Korea.

> [T]he Chinese Army . . . planned attacks to strike from the rear, cutting escape and supply routes and then sending in frontal waves. The basic battle tactic was the *Haichi Shiki*, a V formation into which they allowed opposing troops to move. The Chinese then would close the side of the V while another force moved below the mouth to stop any attempts at escape and to block relief columns.[8]
>
> — *Korea: Untold Story of the War,* by Goulden

That the forward-most American units on Biak had gotten into deep trouble is clear from the battle chronicles. A key step in the *Haichi Shiki* is cutting of any means of escape or reinforcement for the entrapped unit.

> By late afternoon the 3rd Battalion's [of the 162nd Infantry] position was becoming untenable. Japanese mortar and artillery fire increased and enemy patrols cut the coastal road to the rear. Obviously, no further advance could be made until the enemy fire from the East Caves area could be stopped by ground attack from the north, by naval fire from the south, or by artillery fire from emplacements to the east.[9]
>
> — "Approach to the Philippines," by Smith
> U.S. Army Center of Military History, 1996

Figure 3.2: The "Haichi Shiki" Ambush

Overall Flow of the Battle

The initial Biak landing had taken place on 27 May 1944, near the town of Bosnek (while nearby defenders held their fire). They knew the Americans would be heading toward the airfields at Mokmer, Sorido, and Borokoe. When the U.S. attack on Mokmer bogged down, Japanese caves all along the north side of the approach road came to life.

> Patrols from the 162nd Regiment had advanced to about 200 yards of Mokmer airstrip when . . . Japanese opened fire with machine guns and mortars. Limestone caves about 1,200 yards north of Mokmer airstrip [were the source of some fire] . . . . Another line of caves . . . north of the [Mokmer] village, and a third section of caves . . . in . . . the Parai defile [took care of the rest]. . . . These caves . . . [would] hold up the American advance . . . for nearly a month.[10]
>
> — *Warfare History Network,* 30 December 2018

With two U.S. destroyers preparing the way, U.S. infantry supported by Sherman tanks again advanced on the Mokmer

Aerodrome. Any Japanese caves *en route* were cleared or sealed shut with explosives. But, many of their occupants had somehow relocated to the Parai Defile, where they would stymie the advancing 162nd Infantry. It soon became apparent that the airfields could not be taken until all caves in the area had been neutralized.[11] It was then that the hard-pressed GIs realized their adversary had annihilation in mind.

> The 162nd Infantry was hemmed in on three sides and under constant fire from concealed Japanese positions. The men would have to be evacuated from their position if they were not to be annihilated, and there was only one way out—the same amphibious landing craft that brought the troops ashore would have to take them off the beaches.[12]
>
> — *Warfare History Network,* 30 December 2018

Using all available seagoing transport and under the protection of artillery, naval gunfire, and air strikes, the men of the 162nd were taken off Biak's southern beaches during the afternoon of 29 May. "By nightfall, the regiment had been evacuated from Parai Village and landed about 500 yards . . . [farther west] to set up a new position. It had been a near disaster, but the evacuation had succeeded."[13]

With infantry units and six light tanks the next morning, the Japanese then launched a follow-up offensive. Sherman mediums were on hand to handle it, but the 162nd Infantry had other problems. "The men were [being] limited to one canteen of water per day."[14] And tiny groups of Japanese soldiers began shooting at them from all over the place. It was as if the American formation had been infiltrated in broad daylight.

> The caves gave the Japanese a system of natural defenses. . . . They allowed Japanese troops to pop out into the open, fire on unsuspecting Americans, and disappear again.
>
> Some were two and three levels deep. A series of tunnels connected the caves and led to hidden exits. The Japanese "laid ambush after ambush," one soldier remembered.[15]
>
> — *Warfare History Network,* 30 December 2018

The 163rd Regiment joined this drive toward the airfields on the first of June. Mokmer was soon captured, but it would be of no use to U.S. aircraft. Enemy mortars and artillery in nearby cavities kept the field from being repaired. U.S. soldiers with flamethrowers went to clear the cavities. But, their burning fluid couldn't reach far enough into the larger ones. There, satchel charges, hand grenades, and gasoline had to be utilized.[16] (See Map 3.3.)

> In late June the Americans attacked the West Caves, north and west of Mokmer. . . . The caves were finally cleared by the end of the month. Now that they were no longer being harassed by enemy fire, the engineers were able to make Mokmer drome operational. On June 22, the airfield began landing fighters.[17]
>
> — *Warfare History Network,* 30 December 2018

Before long, Col. Riegelman—Gen. MacArthur's chemical officer—showed up to make a study of Biak's caves. Most revealing were the sumps "that might also be the entrance to a succession of chambers and interconnecting galleries with multiple entrances."[18] They provided a glimpse into the travel network that had been carved out of rainwater runoff channels.

On June 20, both Borokoe and Sorido airstrips were taken by American troops. But, about 1,000 Japanese soldiers within the East Caves were still shelling all three bases. So, after the West Caves were cleared, the GIs focused on the East Caves. Most of its threat had been neutralized by 5 July. Any defenders not killed in the assault "slipped back into the jungle."[19]

The Americans—specifically the 163rd Regiment—next approached the Parai Defile and Ibdi Pocket (about 4,000 yards east of the village of Parai). The attack on them went from mid-June well into July. Though over 40,000 mortar and artillery shells had been fired into these enemy bastions during the first two weeks of the assault, their occupants still stiffly resisted.[20]

Near the end of July, all organized resistance on Biak supposedly ended (though some "mopping-up" continued). The island was finally declared secure a month later. From 27 May to 20 August, 400-500 Americans had been killed and 2,000-2,400 wounded, while another 1,000-7,234 GIs had become ill. The extent of Japanese losses widely differed between sources. During the initial battle period, 4700 enemy had been killed, 22 captured, and an unknown number wounded. So had 600 Javanese and British Indian slave laborers been "recovered." But, Sixth U.S. Army would later claim 6,127 Japanese killed or found dead on Biak, plus 462 captured (possibly to include 304 Formosans).[21]

The IJA's Counter-Invasion Scheme

U.S. forces had been privy to a written Japanese defense plan for Biak a full month before the landing. But, that plan may not have addressed what role the subterranean spaces would play. Western bureaucracies run on established procedure. Sub-surface and rearward-moving defenses are largely immune to offensive firepower, and not consistent with America's style of war. They are thus hard for U.S. staff officers to visualize. To this day, U.S. infantry manuals fail to discuss their potential.

> Colonel Kuzume's initial plan for the defense of Biak was published on 27 April, just five days after the Allied landings at Hollandia and Aitape ("2nd Army Opns at Sarmi and Biak" [Rev], pp. 50-56). . . . The bulk of the 2nd Battalion, 222nd Infantry, was to defend the airfields and the coast from Bosnek west to Sorido. The 3rd Battalion (less two companies and some artillery and mortar detachments) was to be held in reserve near the airfields, and the tank company was ordered to assemble near Saba. . . .
>
> . . . [O]n Z Day . . . [the] bulk of the 2nd Battalion, the rest of the naval guard unit, and some naval anti-aircraft organizations were located at the East Caves. Naval headquarters, various naval service units, and the 6th Company, 222nd Infantry, were at the West Caves. Most of the army service units were at Mokmer Drome or disposed along the low ridge north of that field. The bulk of the 3rd Battalion was [at that time] posted at the west end of the same airfield. . . . The tanks . . . were assembled on the terrace north of the eastern end of Mokmer Drome.
>
> At various points along the terrace and low ridge were emplaced a battery of mountain guns, four 120mm naval dual-purpose guns, three or four 3-inch anti-aircraft guns, and a large number of mortars and automatic weapons of

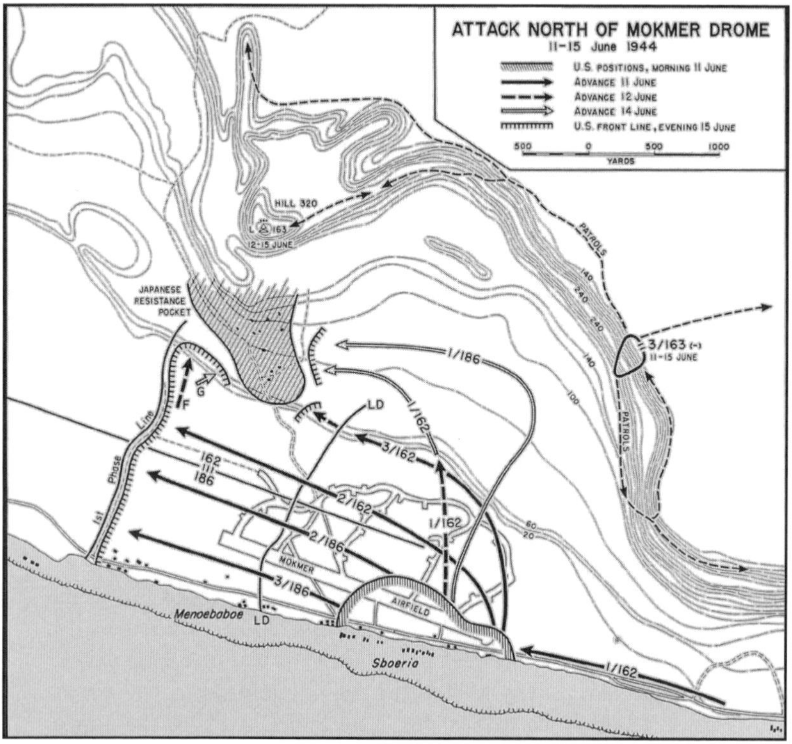

Map 3.3: Attack on Initial Pocket Around "West Caves"
(Source: "The Approach to the Philippines," by Robert Ross Smith, U.S. Army in WWII [Washington, D.C.: Ctr. of Mili. Hist., 1996], map 15.)

all calibers.[22]

— "Approach to the Philippines," by Smith
U.S. Army Center of Military History, 1996

Cave Types on Biak

Just behind Biak's southern beaches lay a steep coral berm. Most of its frontally facing caves had rear entrances on the coast road. (See Figures 3.3 and 3.4.)

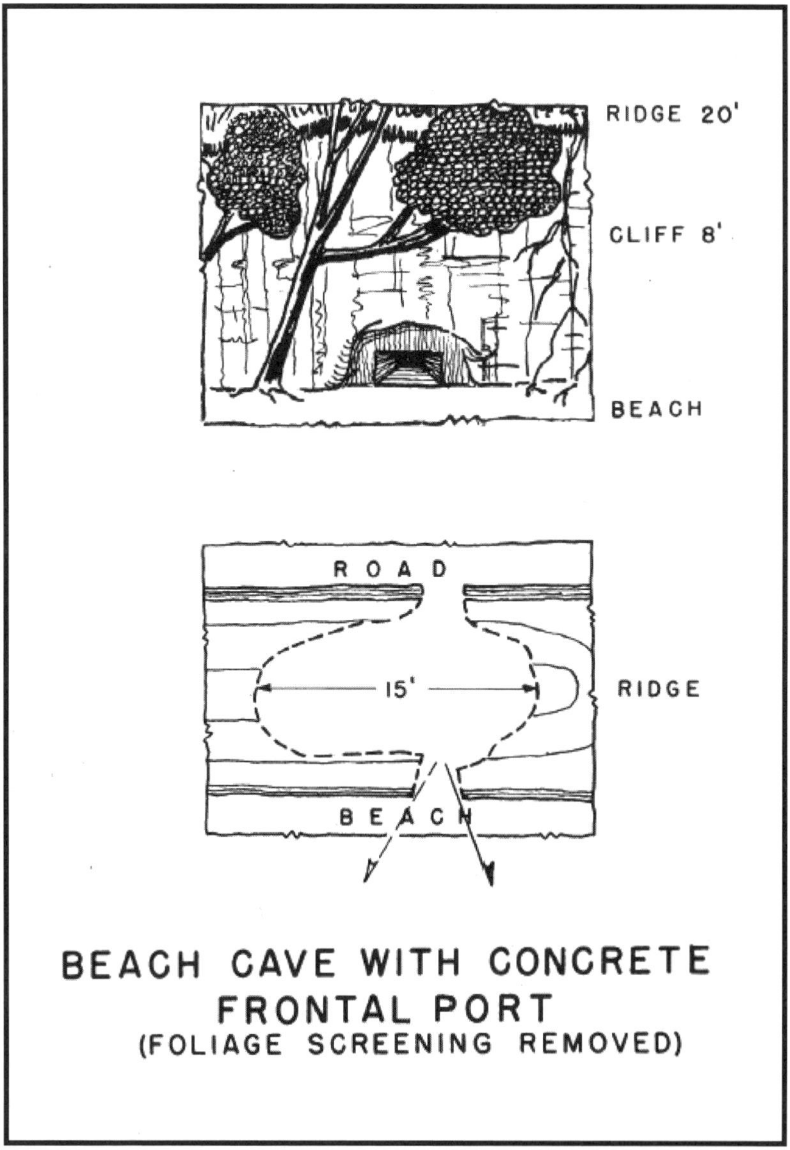

Figure 3.3: Visible Caves Between Beach and Coastal Road
(Source: "History of the Biak Operation 15-27 June 1944," U.S. 41st Army, 1944, Armed Forces Staff College Library copy, p. 45.)

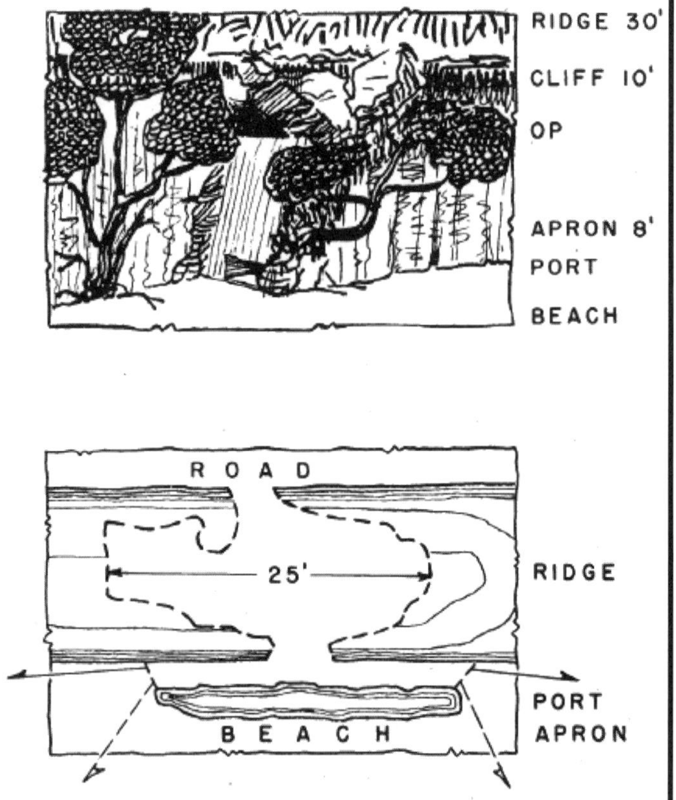

PORTS AND OP AT ENDS OF ALLEY BETWEEN CLIFF AND APRON MASKING CAVE. (FOLIAGE SCREENING REMOVED)

Figure 3.4: Non-Visible Cave Openings Along the Beachfront
(Source: "History of the Biak Operation 15-27 June 1944," U.S. 41st Army, 1944, Armed Forces Staff College Library copy, p. 44.)

The Parai Defile contained cliff-side galleries on the other side of that road. Some 80-feet high, they had a 75° slope of loose debris at their bottom.[23]

At what the Japanese called the West Caves, sumps existed. These oversized sinkholes had gotten their unusual characteristics from once being at sea level. (Look back at Figure 3.1.) They now provided hubs for drinking-water collection and sub-surface travel.

And finally, there were the more common cavities of the West, East, Parai Defile, and Ibdi Pocket cave systems. Formed by the same limestone erosion that had produced the sumps, most were connected by runoff channels to a deeper set of chambers. (See Map 3.4.)

Protection of the Major Cave Sets

Biak's defenders didn't build as many coconut log pillboxes as earlier in the war. At the center of their building-block strongpoints were caves with cement gun ports. Some ports had steel covers. Other tiny perimeters had been established around buried cubicles of reinforced concrete. (Look back at Figures 3.1 and 3.2.) But, most encircling dugouts were connected by slit-trench to the central bomb shelter. That type of defiladed passageway was easier to build at the somewhat elevated West Caves.[24] Trenches also connected most parts of the Ibdi Pocket strongpoint matrix.

> In this [Ibdi Pocket] area the Japanese had constructed defensive installations varying in type from heavily constructed pillboxes of log, coral, and concrete to hasty trenches in the coral and brush.[25]
>
> — "Approach to the Philippines," by Smith
> U.S. Army Center of Military History, 1996

Every Biak cave system had been surrounded by tiny man-made forts, each with its own machinegun. As such, its surface fortifications resembled the Bairoko bunker belts with some trenching added.

> The West Caves were actually three large sumps, or depressions in the ground, which were connected by

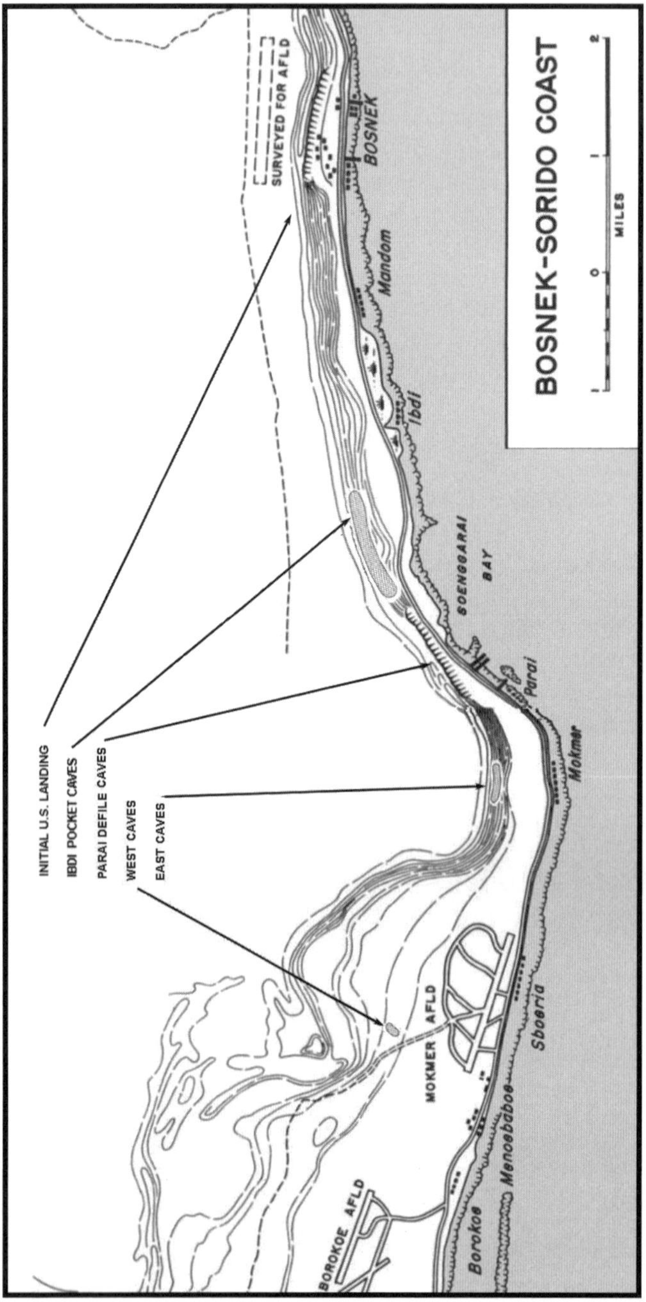

Map 3.4: Japanese Caves in Relation to Ridgeline and Road
(Source: "The Approach to the Philippines," by Robert Ross Smith, U.S. Army in WWII [Washington, D.C.: Ctr. of Mili. Hist., 1996], map 14.)

> underground tunnels and caverns. The caves were ringed with pillboxes, bunkers, and foxholes, and an extensive system of coral and log emplacements was built along the spur ridge above Mokmer Drome. Biak naval headquarters was originally located in the West Caves, which could shelter 1,000 men.[26] (Figure 3.1 is thus a depiction of the West Caves.)
>
> — "Approach to the Philippines," by Smith
> U.S. Army Center of Military History, 1996

Farther up the ridge, the East Caves had entrance caverns instead of entrance sumps. From those caverns, the enemy sporadically shelled all three air bases. Most of the East Caves had been neutralized by July 5. That's when the rest of its 1000-man contingent "slipped away into the jungle" (or by some other avenue to Parai Defile).[27]

> East Caves. Behind [the village of] Mokmer the ridge rose to a height of 240 feet. . . . About three quarters of the way to the top was a flat ledge from which two large caverns, similar to those in the West Caves area, could be entered. The Japanese constructed pillboxes on the ridge both below and above the ledge, and in the caverns they emplaced mortars, 20mm guns, and heavy machine guns.[28]
>
> — "Approach to the Philippines," by Smith
> U.S. Army Center of Military History, 1996

The Parai Defile caves were protected by cement pillboxes atop the first-terrace cliff and rear entrances to shoreline berm cavities.[29]

> Parai Defile. [A] level series of intermittently connected cavities about 4 to 8 feet in height and 3 to 6 feet in depth. Stalactite-stalagmite unions and hard limestone formations occasionally interrupted continuous passage. In several instances, tunnels arched [around] or ran behind these obstructions. . . . The face of these cliffs rose another 80 feet above the galleries. . . . Above the face of the cliff, there were suitable M.G. [machinegun] emplacements.[30]
>
> — "History of the Biak Operation," 41st Division

The Ibdi Pocket caves, to which the airfield defenders finally withdrew, were those nearest the original landing site. Here, the landscape permitted a grid pattern of interlocking machinegun fire that would badly hurt any intruder. (The upper slopes of Umurbrogol Mountain at Peleliu would host a similar array of mutually supporting caves, and the Amphitheater at Iwo Jima have its man-made equivalent.[31])

> [T]he Ibdi Pocket. The terrain in the area was a series of knife-like east-west ridges separated by depressions and crevices up to fifty feet deep. These ridges were connected in places by cross-ridges, and the entire area was covered with thick rain forest and dense jungle undergrowth which had found a foothold in the coral. Pillboxes of coral and logs, hasty emplacements of the same materials, small caves and crevices, and foxholes at the bases of large trees were all utilized by the enemy to defend the area.[32]
>
> — "Approach to the Philippines," by Smith
> U.S. Army Center of Military History, 1996

At Ibdi, Japanese engineers had found the perfect location for a last stand. After severely attriting any attackers, its rear party could easily slip away. Man-made tunnels had existed at the West Caves and Parai Defile. Another big one at Ibdi Pocket would provide access to many of its composite cavities.

> The main enemy defensive position [at the Ibdi Pocket] was . . . in a large cylindrical hole with caves leading off from its side.[33]
>
> — "History of the Biak Operation," 41st Division

There is no way of telling for sure if all four of those Japanese cave sets sat above a subterranean stream following the base of Mokmer terrace. (Look back at Map 3.4.) But, large numbers of enemy troops had been too easily circumventing their U.S. cordons.

Had the Four Cave Sets Been Connected Below Ground

The southern end of Biak Island had been covered with sub-surface drainage canals. Because caverns at the East Caves had performed the same function as West Cave sumps,[34] a waterway beneath Mokmer Ridge had likely flowed between the two cave complexes.

Just as stalactites and stalagmites were easy to remove, so too could the diameter of the eastern end of that Mokmer Ridge waterway be enlarged. The resulting travel conduit would have allowed West and East Cave defenders to secretly move—through Parai Defile—all the way back to the Ibdi Pocket. (Look back at Map 3.4.) Besides an immense *Haichi Shiki* along the coastal road, so doing would have made possible the destruction of America's resupply facility.

Nipponese Actions After the Road Ambush

Within the urban-warfare procedures emerging from Europe, the ambush of an enemy attack column was often followed by the defenders' counterattack.

> The forward elements [of the 162nd Infantry] had gotten within 200 yards of Mokmer airfield when the Japanese defenders suddenly opened fire from the East Caves and Ibai [Ibdi] Pocket, ambushing the American column. Elements of 2/222 Regiment attacked from the west and north and soon had 3/162 Regiment trapped in a perimeter 200 yards deep and 500 yards long. . . [Parts of the U.S.] Regiment attempted to outflank the Japanese by advancing along the ridge from Parai but were stopped by fire from East Caves. A platoon of Sherman tanks helped drive back an attack supported by Japanese tanks, but 1st and 2nd Battalions [of 162nd Infantry] were unable to break through to relieve 3rd Battalion. Supplies had to be brought in, and casualties evacuated, by landing craft. The naval support officer with 162 Regiment was killed around noon, which greatly hindered cooperation with destroyers offshore. At 1600 [Gen.] Fuller ordered 3rd Battalion to pull back behind 2nd Battalion's lines. Four of the Shermans covered the withdrawal, which was completed by 1900.
>
> Destroyer gunfire was unsuccessful in driving the Japanese back during the night, and the Japanese moved

1st Battalion, 222nd Regiment to West Caves from its bivouacs to the north and prepared to throw all of 2nd and 3rd Battalions against the Americans the next morning. The counterattack was supported by light tanks, but these were completely outclassed by the Shermans, which knocked out seven of the Japanese tanks while the Japanese infantry took heavy casualties. However, by noon the Japanese had cut the coastal road at Parai and it was clear that 162nd Regiment was in an untenable position. The American tanks were able to cover the evacuation of 2nd Battalion in LVTs [Landing Vehicles Tracked] while the remainder of 162nd Regiment cleared the roadblock at Parai and moved back to Ibdi (136.190E 1.178S).[35]

— *The Pacific War Online Encyclopedia*, n.d.

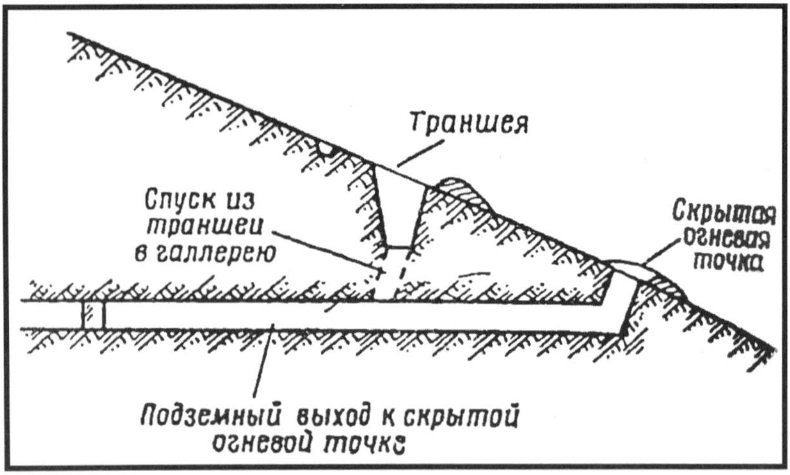

Figure 3.5: Subterranean Escapes Were Common in the East
(Source: *Voina v Koree: 1950-1953,* by A.A. Kuryacheba, © 2002 by Polygon Publishers, p. 614.)

Not only could the Biak caves provide a full measure of surprise fire and static defense, but also of counterattacking personnel. There had been 900-1000 Japanese soldiers at the West Caves,[36] and 1000 more at the East Caves.[37] There were also "large garrisons" in the caves above Mokmer Village (possibly to include Pari Defile) and at the Ibdi Pocket.[38] If all those enemy soldiers had come up out of the ground at once, the upper portion of that leaning "V" would have irresistibly swung shut against Biak's shoreline.

Final Reduction of the Ibdi Pocket

The sub-surface labyrinth below Ibdi Pocket had been sizable, for the final massive bombardment did little good. Only after 200 Japanese troops had exfiltrated the cordon, did all remaining holdouts supposedly succumb to 1000-pound bombs. In actuality, they may have simply returned below ground to one of the other cave complexes, each with its own exits to the jungle. Or the Ibdi rear party could have snuck beneath its own U.S. cordon through a prearranged escape tunnel. (See Figure 3.4.)

> The next day [25 June 1944] more units were brought forward until, by noon, almost all of the 1st and 2nd Battalions, 163rd Infantry, were in action against the pocket. . . . By the end of the day . . . the two battalions had compressed the Japanese into an area approximately 600 yards square. Attacks were continued on the 28th, but little more progress was made. . . .
>
> The 28th of June marked the end of large-scale infantry action at the Ibdi Pocket. Instead, extensive patrolling was undertaken in order to pin-point targets for artillery, mortar, and air bombardments. . . . [O]n 9 July Allied Air Forces P-39's and P-40's dive-bombed and strafed the pocket area.
>
> . . . . From the 21st of June through the morning of 10 July the 146th Field Artillery Battalion alone had thrown about 20,000 105mm. shells into the Ibdi Pocket, and at least as many rounds had been fired by 4.2-inch, 60mm and 81mm mortars.
>
> Beginning on the 11th, the 3rd Battalion, 163rd Infantry, steadily compressed the remaining Japanese into a smaller

and smaller area. Bazookas and flame throwers were used to good purpose in close-in operations against bunkers, pillboxes, improved caverns, and small caves. Tanks fired their 75mm weapons at enemy positions now visible from the northern side of the main ridge, and on the 12th 155mm howitzers of Battery C, 947th Field Artillery Battalion, were brought forward to fire on the pocket. The next day about 200 Japanese, in four separate groups, sneaked through a net of ambushes and patrol bases set up by the 163rd Infantry on the flats north of the main ridge. Despite this exodus, which must have represented most of the . . . able-bodied members of the 3rd Battalion, 222nd Infantry, . . . American infantry moving back into the pocket on the 15th could not report that the going was much easier.

Artillery bombardment was resumed. . . . From 10 to 20 July the 146th Field Artillery Battalion threw in over 4,000 shells and Battery C of the 947th, fired 2,000 rounds into the pocket. On 22 July eight B-24's swept over the pocket in three waves, dropping sixty-four 1,000-pound bombs. Artillery units followed with 1,000 rounds of 105mm and 275 rounds of 155mm ammunition, while the 163rd Infantry's 81mm mortars threw uncounted shells into the pocket. These bombardments halted at 0945 on the 22nd and five minutes later two under-strength companies of the 3rd Battalion, 163rd Infantry, started moving toward the enemy's remaining defensive installations. The 1,000-pound bombs had paralyzed the Japanese, and the two American units encountered almost no opposition. At 1835 the 3rd Battalion commander reported that the Ibdi Pocket was cleared of all organized resistance.[39]

— "Approach to the Philippines," by Smith
U.S. Army Center of Military History, 1996

Special Snipers Had Protected Biak's Fortified Areas

As the Marine Raiders neared the Bairoko bunker belts in July of 1943, they had encountered treetop rifle shots and *Nambu* fire from outpost foxholes.[40] Something similar would occur at the Biak Sumps.

> Jap snipers had been unusually active from their well concealed positions in crevices and recesses in the ridges around the Sumps.[41]
>
> — "History of the Biak Operation," 41st Division

The precise role of those Biak snipers not clear. When American Army troops finally assaulted the Sumps before dawn on the 26th of June 1944, they found the forward-most enemy emplacements unmanned, but foe snipers still protecting the first row of bunkers.[42]

> They [the Biak attackers] later contacted a series of enemy emplacements well organized in depth and protected by heavy sniper fire.[43]
>
> — "History of the Biak Operation," 41st Division

Those Japanese snipers had been performing more than just forward sentry post duty. Their shooting had provided a deception of sorts, because MLRs normally identify themselves through automatic-weapons fire. Large urban areas in Europe were also being blanketed with snipers to keep their defense lines from being discovered.[44]

Riflemen at various distances from a Biak Sump could have more easily ascertained the route of an approaching attacker, and ahead-situated *Nambu* teams drawn U.S. attention away from the bunker line.

Many years later, North Korea's standard strongpoint would dictate an ambush to its front—probably to confuse enemy soldiers as to its exact whereabouts.[45] That's why so many U.S. attackers of the Biak Sumps had been ambushed *en route.*[46]

On Guadalcanal in 1942, Nipponese sniper/sentries hid in tree trunks and treetops.[47] Some had even functioned as forward observers for Japanese artillery.[48] But on Biak, such personnel would—for the first time—have access to a sub-surface dimension. By enhancing their safety, this broadened their use.

Biak's Refinements to a Bairoko-Like Defense

By the summer of 1944, Tokyo was clearly losing the war in the

Pacific. As had the Germans in 1917 (by replacing the continuous trenchline with a belt of squad strongpoints [49]), the WWII Japanese were refining their new defense strategy.

Bairoko had produced offset rows of roofed bunkers that, while mutually supporting and covered by friendly mortar fire, would discourage the average attacker.

With Biak came the opportunity to combine a labyrinth of subterranean passageways with surface strongpoint matrices. Together, they produced a "soft" and rearward-moving defense in depth—one that would bend but never break. Because such an arrangement took advantage the defenders' remaining assets, it made the gradually weakened intruder even more susceptible to counterattack. Besides shifting unused personnel and ordnance rearwards, it allowed for the following: (1) a heavier shelling of one's own positions during any withdrawal; (2) spoiling attacks from overrun sub-surface spaces; (3) the covert advance of counterattacking forces; and (4) an easy escape for any encircled element.

At Biak, those Japanese snipers may have also guarded the entrances to sub-surface passageways (like the VC during the Vietnam War [50]). There had been tunnel-connected strongpoint rings around the West Caves, East Caves, Parai Defile, and Ibdi Pocket. After the Ibdi Pocket cordon was tightened, its holdouts could have escaped at night through a previous emplacement.

This sophisticated a Japanese defense became important in the summer of 1944 because many islands to the north of Biak consisted of coral limestone. Those of any height were already laced with subterranean passageways from erosion. Others were composed of volcanic rock that was even easier to excavate. So, if enough engineers and slave labor could be employed on the most likely U.S. targets, the Americans' might be forced into an "armistice" (as in Korea nine years later).

Conventional Wisdom After the Battles of Bairoko and Biak

At Biak, the U.S. Army had done what it does best—systematically grind down all opposition. Though the entire enemy garrison had been supposedly killed, the number of corpses would never substantiate that claim.

Not until Okinawa's infamous Shuri Line would America's steamroller approach to an MLR prove fallible.[51] There again, enemy bastions with sub-surface chambers would receive enough diesel fuel and TNT (Trinitrotoluene) to temporarily silence them, but then come back to life.

All the while, the exorbitant expenditure of American ordnance would further the myth that modern defense belts cannot be taken any other way. Without tanks, artillery, and air power, U.S. infantrymen would never breach one. What the 4th Raiders had done at Bairoko without any heavy weaponry would not be remembered for long. It was to help the 4th Marine Regiment on Guam and Okinawa, but not make it into the Corps' "institutional knowledge" after the war.[52]

PART TWO

Then Peleliu Was Encountered

"AND WHEN HE GETS TO HEAVEN, SAINT PETER HE WILL TELL:
'ANOTHER MARINE REPORTING, SIR—I'VE SERVED MY TIME IN HELL!' "

(Source: Possibly a revised version of "Our Hitch in Hell," as found in the book "Mud and Stars - An Anthology of World War Songs and Poetry," by Dorothea York © 1931.)

4 Next Stepping Stone for U.S. in the Pacific

- Had Marine leaders foreseen all obstacles to the invasion?
- What would the landing force face on its first day ashore?

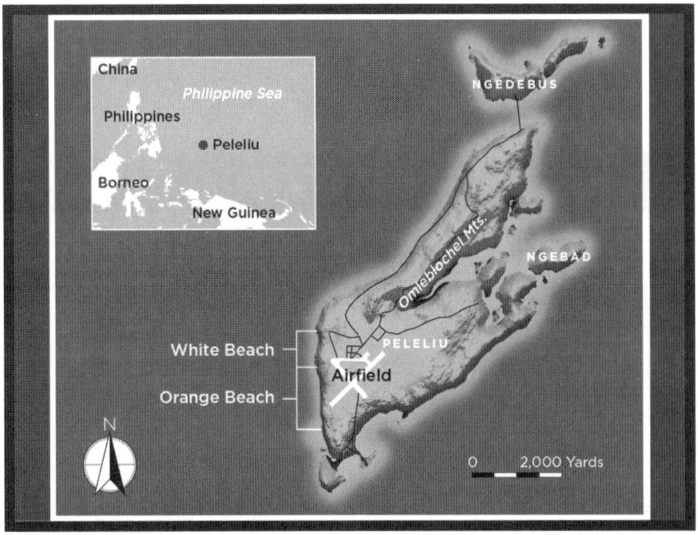

(Source: Courtesy of Ken Feisel, "Archaeology Magazine," from https://www.archaeology.org/issues/344-1907/features/7712-peleliu-wwii-pacific-battle, ©)

Peleliu's Particulars

Peleliu was no tourist hangout. Six-miles long and two-miles wide, it had only five-square miles of solid ground. The rest was tidal lagoon and mangrove swamp—infested by huge seagoing crocodiles. That means sitting alone along its shore at night to escape the oppressive heat was risky.

This "South Sea" island's bottom end was flat, partly covered by small bushes or trees, and host to a good-sized airfield. The beaches to the west of that field were best suited for an amphibious landing. A few coconut palms grew in the brush along them.[1]

Figure 4.1: The Palaus Under German Rule in 1908
(Source: "Palau under German rule," painting by Rudolf Hellgrewe, 1908, Wikimedia Commons, public domain, with image designator "Palau-Inseln.jpg.")

Just up the island from the aerodrome was Umurbrogol (a.k.a. Omlebrochel) Mountain and more northeasterly running high ground. Within this elongated hill-mass was a combination of rugged raised coral, limestone peaks, and ridges. Those ridges were, in turn, honeycombed with outcroppings, gorges, crags, sinkholes, and caves.[2] Umurbrogol Mountain itself had several pinnacles that rose some 300 feet above sea level and steep cliffs of 40-60 feet in height. Some slopes of this mountain contained lateral cracks that had been hidden by the vegetation. Prior to landing, U.S. intelligence would fail to realize just how broken that hillside really was or what had been done to fortify it.[3]

Peleliu was one of the Palau group of islands some 430 miles east of Mindanao. Control of that group had passed from Germany to Japan in 1914. (See Figure 4.1.) Peleliu would be one of the tiny specks in the Pacific that Japan fortified in its "second line of defense."[4] By September of 1944, this tropical isle was being occupied by about 11,000 Nipponese soldiers of the 14th Infantry Division, 1000 or so naval personnel, and 1500 Korean and Okinawan laborers.[5] Col. Kunio Nakagawa, commander of the Division's 2nd

Regiment, had supervised many of the defense preparations. Its hundreds of limestone caves were now the nucleus of a formidable fortress.[6] (See Maps 4.1 and 4.2.)

Despite Peleliu's imposing landscape, U.S. warplanners had estimated its capture in three to five days. Then, the struggle would last for two and a half months and be the fiercest to date in the Pacific.

What made Peleliu so difficult to capture will be the objective of this study. That answer may significantly impact the future. Islands of the same latitude and composition guard the entrance to the Straits of Malacca. They may be soon seized by the PRC to secure a principal trade route for wartime materiel.

Initial Problems with the Peleliu Landing

When fully reinforced, 1st Marine Division consisted of 16,478 brave souls. After a three-day naval bombardment, some 5000 of them made the initial landing on Peleliu. Most were shocked by what the Japanese had waiting. (See Figure 4.2.) Not only were its defenders firing down from Umurbrogol Mountain, but also from both flanks at any point along the 2,200-yard beachhead.[7] It was this enfilade fire at the waterline that did most of the damage.

Figure 4.2: Peleliu-Bound Leatherneck in 1944

(Source: "Bloodly Beaches: The Marines at Peleliu," by B.Gen. G.D. Gayle, Marines in WWII Commemorative Series, p. 3, image designator "Page_05_Image_0003.jpg.")

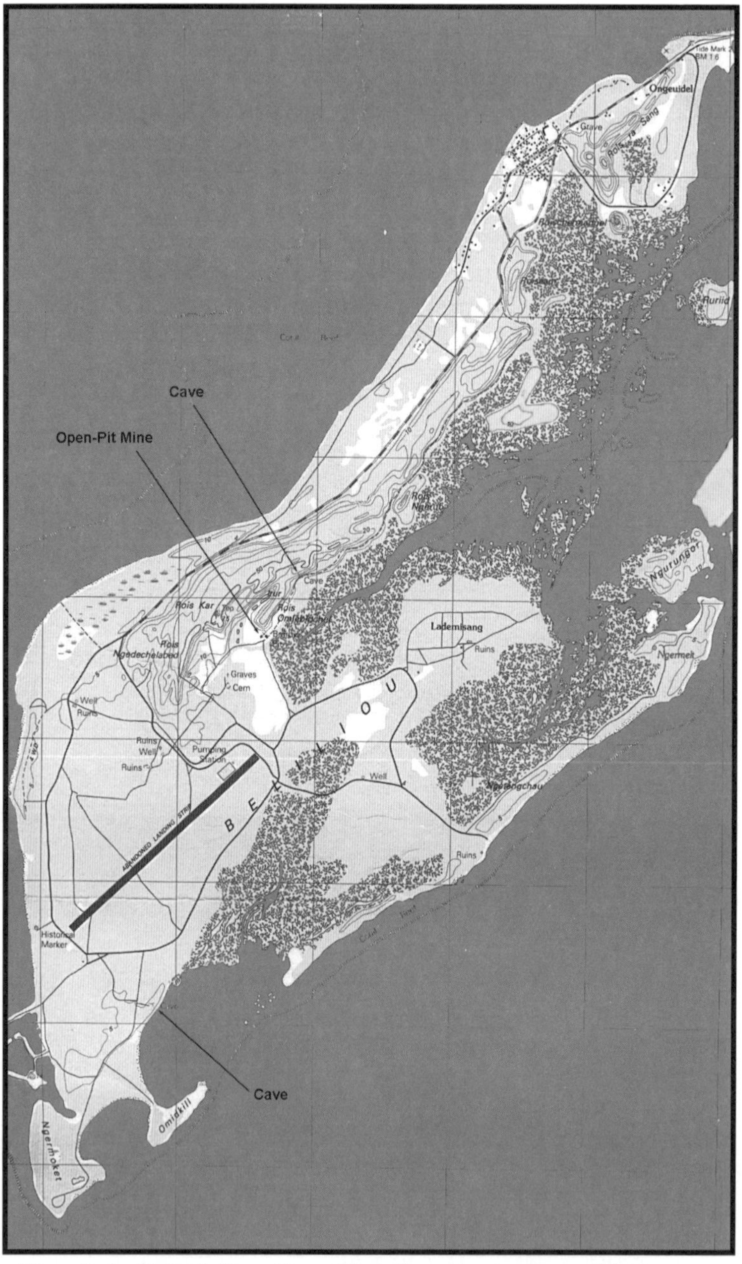

Map 4.1: Vegetation Map Showing Swamps and Lagoons

(Source: "Map of Peleliu,"Perry Castaneda Collection, from https://upload.wikimedia.org/wikipedia/commons/2/2d/Peleliu_map.jpg,U.S,Geological Survey,public domain.)

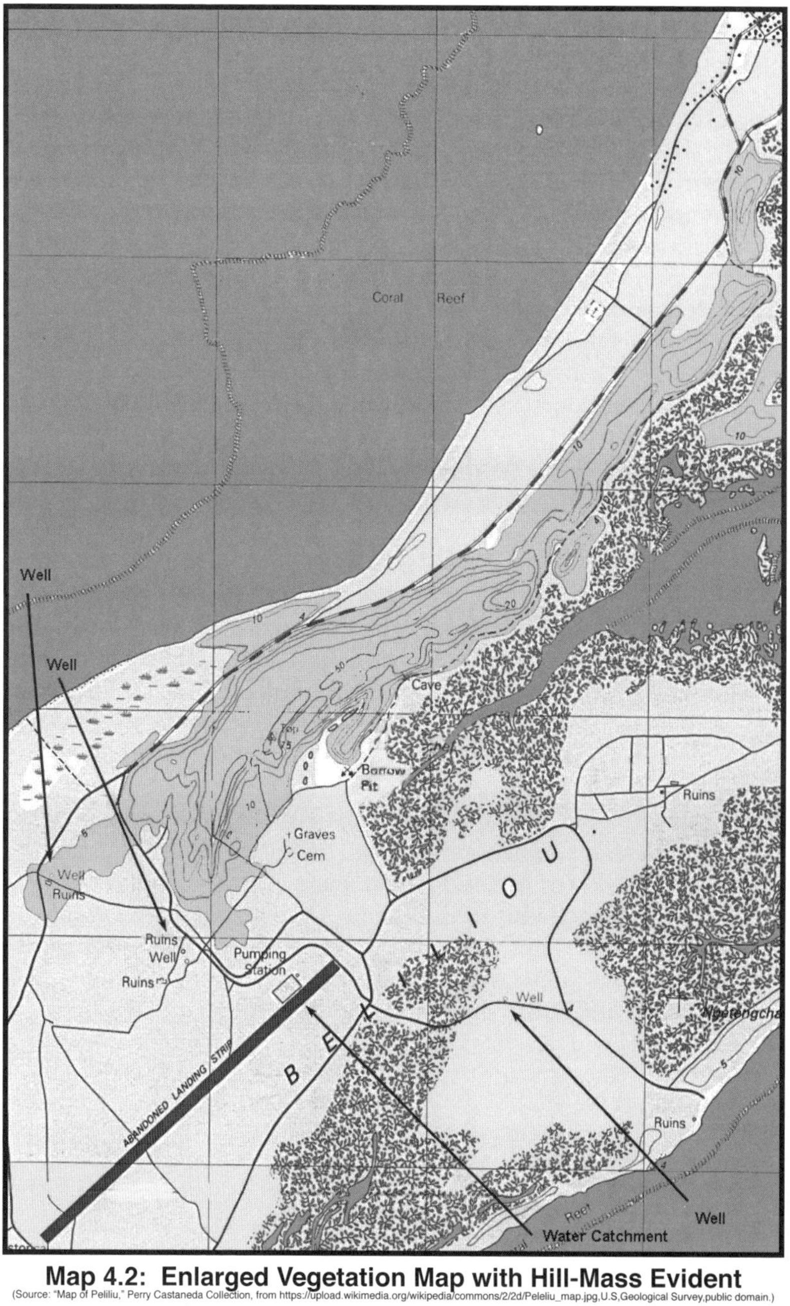

Map 4.2: Enlarged Vegetation Map with Hill-Mass Evident

(Source: "Map of Peliliu," Perry Castaneda Collection, from https://upload.wikimedia.org/wikipedia/commons/2/2d/Peleliu_map.jpg,U.S,Geological Survey,public domain.)

> On D-Day 15 September 1944, five infantry battalions of the 1st Marine Division's 1st, 5th, and 7th Marines, in amphibian tractors (LVTs) . . . lumbered across 600-800 yards of coral reef . . . and toward . . . landing beaches. . . .
>
> Just ahead of the first wave of troops carrying LVTs was a wave of armored amphibian tractors (LVTAs) mounting 75mm howitzers. . . . And just ahead of the armored tractors, as the naval gunfire lifted toward deeper targets, flew a line of U.S. Navy fighter aircraft, strafing . . . along the length of the beach defenses. . . .
>
> . . . [A]s the leading waves neared the beaches, the LVTs were hit by heavy enfilading artillery and anti-boat gun fire coming from concealed bunkers on north and south flanking points.[8]
>
> — *The Assault on Peleliu,* by Hough
> History Division, HQMC, 1950

After the initial wave of LVT(A)s—each mounting a 75mm howitzer or 37mm antitank gun—would come eight waves of Marine infantrymen in LVTs (each designed to hold 20 men). Then, the Division's organic artillery support—the 11th Marines 75mm and 105mm howitzers—followed in LVT-4's and DUKWs.[9] But, there had been a price to be paid for all this activity. And here's how the Corps' most beloved officer remembers his part of the landing.

> "I went up and over the side [of the LVT] as fast as I could scramble and ran like hell at least 25 yards before I hit the beach, flat down. When I looked back to the Amtrak I saw four or five shells hit it all at once. A few men were killed, getting out too slow, but most of them were saved because they got out before we stopped moving. . . . I looked down the beach and saw a mess–every damn Amtrak *[sic]* in our wave had been destroyed in the water by the enemy, or shot to pieces the minute it landed." (Davis 1962:217, Denfeld 1988:15).[10]
>
> — Col. Lewis Puller, after landing on White Beach

The Enemy's Overall Plan for Peleliu

While Peleliu's broken terrain offered many 2GW opportunities for "attriting" a Western intruder, it was also well suited for a 3GW-like campaign. Such a campaign would partially dodge the attacker's thrusts to more easily get at his command and logistics networks.

By 1944, this new way of operating had already been successfully tried by the Soviet defenders of Stalingrad.[11] In early 1945 at Iwo Jima, it would help the besieged Japanese to conserve both manpower and ordnance through: (1) only defending where the invader was about to advance; (2) shifting unused assets rearward after every engagement; and (3) then resisting from farther back.[12] Iwo's occupiers called it "retreat combat,"[13] as had their German allies. Its eventual goal, of course, was to make the intruder more susceptible to counterattack. But, when combined with guerrilla warfare on Peleliu, it might also delay a U.S. invasion of the homeland.

As at Biak, another aspect of retreat combat on Peleliu was drawing the invader into preplanned kill zones. Its specifics would vary. But, so doing was consistent with the core stratagems President Kennedy would later attribute to Unconventional Warfare (UW)—infiltration and ambush.[14] Many of the spoiling attacks on Peleliu would be directed—from below ground—at the Americans' rear-area support apparatus. That's why UW is now considered a subset of 3GW.

The Palau Defenders' New Scheme of Maneuver

On Peleliu, the IJA was to conduct a mixture of what it had done at Bairoko and Biak. The Battle for Bairoko had produced a more deadly "defense in depth"—offset rows of roofed bunkers that could be dusted with mortar fire. Then, Biak was the first full test of the IJA's new island protection plan—no beach resistance or human-wave attacks. Its garrison had instead tried—with an inland ambush—to drive the invaders back out to sea. Then, they had funneled all unused manpower and ordnance back toward the original landing site through four sets of caves. That there had been U.S. cordons around all four suggests an underground passageway between them.

During the battle for Peleliu, Col. Nakagawa was to put the finishing touches on the IJA version of "retreat combat." He would marry his man-made emplacements to rainwater erosion cavities and run-off canals. By so doing, he was able to create a mutually supporting honeycomb of bunkers, caves, and subterranean passageways that would be hard to penetrate. (See Map 4.3.) Some of the bigger caverns served as a staging area for his "Reserves" and possibly tanks. (See Map 4.4.) Then, any hard-pressed frontline fighters could fall back to join those Reserves. And, as the Japanese perimeter gradually shrank in size, that growing manpower pool made possible big spoiling attacks. Some would be launched from below ground at places where many members were unlikely. (Map 4.4 discloses those locations.)

Only a Few Japanese Bastions Would Attempt to Holdout

Nakagawa's protective formations were mostly on Umurbrogol Mountain at the south end of Peleliu's central hill-mass. Overlooking the strategically important airfield, that mountain contained many of the island's 500 limestone caves. Some next to mine shafts had been turned into gun parks. To them, Japanese engineers added sliding steel doors and multiple openings for artillery and machine-guns. Other cave entrances had been slanted for protection against grenade and flamethrower attacks. Then, the most important caves and their supporting bunkers were connected by a vast tunnel system. The resulting movement/communication network allowed the Japanese to evacuate or reoccupy positions as needed, and to circumvent attacker cordons.[15]

What U.S. Intelligence Knew About the Enemy Strategy

Within the appendix to an official U.S. chronicle are excerpts from a captured Japanese document entitled "Palau Group Training for Victory." This document was issued by an IJA Group Headquarters in July of 1944 (two months before the Peleliu landing). It offers the first written insights into Japan's new island resistance regimen.

> "The ultimate goal of this training is to minimize our losses in the severe enemy pre-landing naval and aerial bombardment. . . .
>
> Compared with our "short sword" the Americans have

a "long spear" with which to do battle. . . . We must first resolutely penetrate to the enemy and then we shall display our short swords. . . .

. . . [E]ven if we die delivering our territories into the hands of the enemy it will not necessarily contribute to the opening of a new (and favorable) phase of the war. . . .

We must preserve personnel and ordnance—psychological and material strength. . . .

. . . [I]n addition to attempting to destroy and confuse enemy landing craft by means of strongly prepared beach positions [hidden bunker complexes], to lead the enemy to confusion and destruction by concentrating rapidly disappearing fire power from our strongpoints remaining in his midst, . . . and to carry out strong counterattacks from previously planned and prepared positions in order to destroy the enemy that has landed by dawn of the next day.

In order to do this it will be necessary to carry out this counterattack with men who have been withdrawn from non-active fronts or else to use as a nucleus well-selected reservists who have been previously withdrawn."[16]

— *The Assault on Peleliu,* by Hough
History Division, HQMC, 1950

The same Japanese document alludes to defenders somehow withdrawing from beachfront strongpoints to later contribute to the battle effort. To maintain enough manpower for an extended struggle, it also discourages honor-saving *banzai* attacks. Still a mystery is how overrun strongpoints could be safely exited. The comment that "mere dying is not enough" confirms the intention to conduct Guerrilla Warfare (GW). A participant in the Battle for Luzon would be later ordered to so.[17] Near the end of the Peleliu struggle, there had been other indicators of GW.

With capture of the high ground immediately dominating the airfield, the campaign degenerated into what was in effect guerrila *[sic]* warfare of a type peculiar to no people other than the Japanese. The defenders' potential for tactical action on any effective scale had been destroyed, and they were cut off from support and reinforcement. From that point onward, they killed solely for the sake of killing,

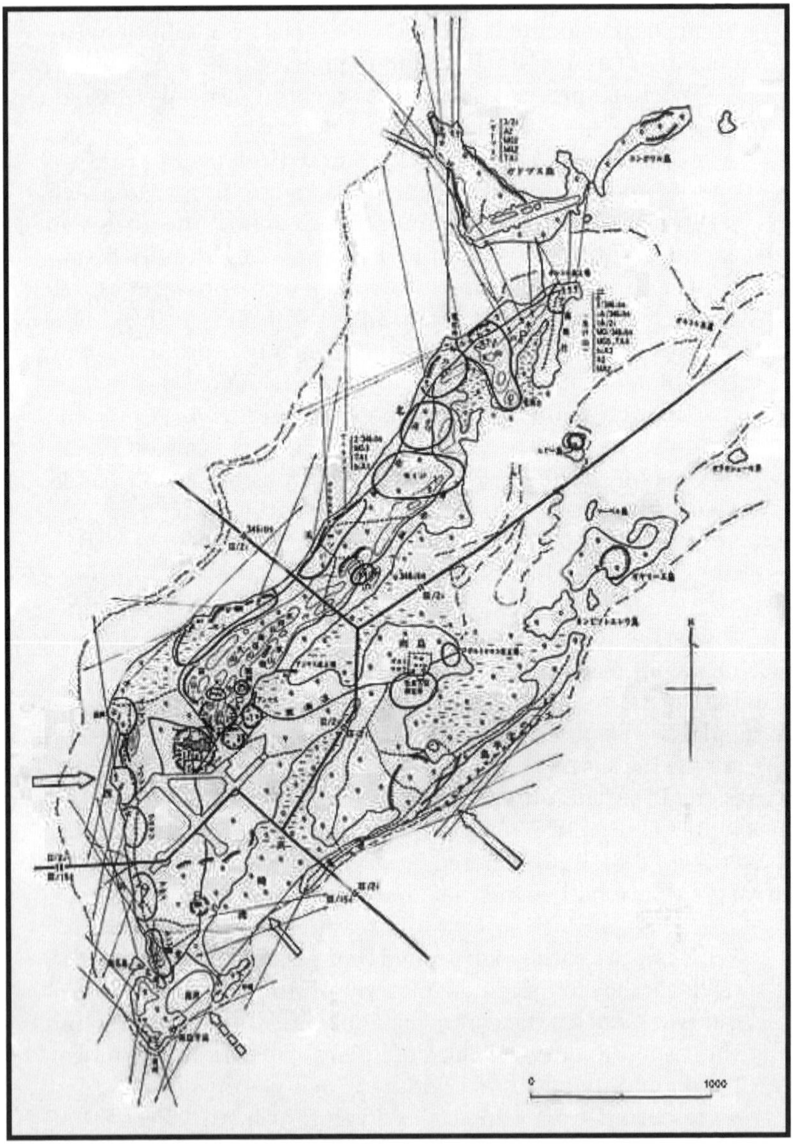

Map 4.3: Captured Japanese Map of Peleliu's Protection Plan

(Source: Japanese Map of Peleliu, History of War website, possibly from captured 31st Army files, at this url: http://www.historyofwar.org/Pictures/pictures_peleliu12.html.)

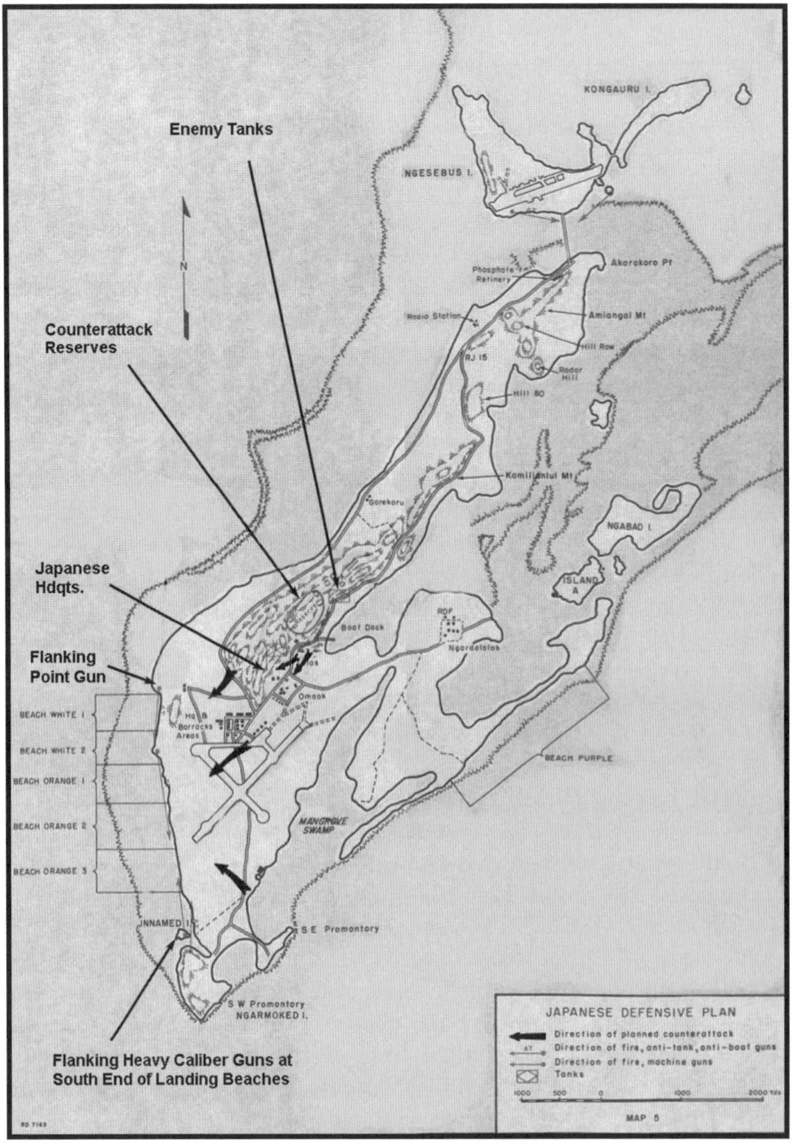

Map 4.4: U.S. Assessment of Japanese Defenses
(Source: "The Assault on Peleliu," by Maj. Frank R. Hough USMCR [Washington, D.C. Historical Div., HQMC, 1950], map 5.)

without hope and without higher purpose.[18]
— *The Assault on Peleliu,* by Hough
History Division, HQMC, 1950

The Enemy's Order of Battle

The 1996 account shows the Japanese forces on Peleliu (to include Ngesebus) at the time of the invasion to consist of: (1) 5388 Army soldiers [14th Division and 53rd Independent Infantry Brigade]; (2) 800-1000 Navy combatants; (3) 2,000-2,200 Navy labor personnel [204th and 214th Construction Battalions, plus elements of the 43rd and 235th]; and (4) 2220 Navy airbase personnel.[19] That makes a grand total of 10,400 to 10,800.

The U.S. National Park Service reports 10,500, of whom 6,500 were combat troops and 4,000 naval personnel. Other researchers say 11,000 Japanese military men and as many as 3,000 Korean and Okinawan forced laborers.[20]

That same 1996 account talks about 25,000 more Japanese troops on the northern Palau island of Babelthuap. Within that number were 1,349 amphibious troops (Japanese Marines). Most shocking is a National Park Service claim that "[t]he Japanese military presence in the Palaus also included three Construction Battalions made up of nearly 10,000 forced labourers."[21] Had the additional thousands of Koreans and Okinawans been present on Peleliu, the way the battle was waged would merit further scrutiny.

Cost of Taking Peleliu

The assault on Peleliu began on 15 September 1944 and did not end until 27 November 1944. There would be some guerrilla activity until February 1945.[22] The casualty figures differ. A civilian study in 2004 puts the American losses at 9,740, with 1790 of those killed in action. It also shows the Japanese sacrifice at 10,695 men, with almost all being fatalities.[23] A documentary film in 2007 claims that 1200 U.S. troops perished, and another 5274 became maimed or missing.[24] But, the initial chronicle indicates the U.S. Army casualties had yet to be added to those totals.

The [Stalemate II] operation was a costly one by any

> standards. Latest . . . checkup gives total Marine Corps casualty figures of 6,526, of whom 1,252 are listed as dead (killed in action, died of wounds, and missing presumed dead). . . . In addition, the 81st Infantry Division, Reinforced, lost 1,393 officers and [enlisted] men on Peleliu, [with] 208 killed in action.[25]
>
> — *The Assault on Peleliu,* by Hough
> History Division, HQMC, 1950

By any measure, it had been a fiercely fought confrontation. Within its specifics should be much of continuing value to U.S. forces.

5 The Initial Ambush at Peleliu

- Why did so many Marines get hurt right away on Peleliu?
- What had the invasion planners failed to notice?

The full length of Peleliu's narrow beach was a death trap.

(Source: "Bloody Beaches: The Marines at Peleliu," by B.Gen. G.D. Gayle, Marines in WWII Commemorative Series, p. 9, image designator "Page_11_Image_0002.jpg.")

The Unimaginable Was About to Happen

At Peleliu, the Americans had intended to land 4,500 assault troops during the first 19 minutes of the invasion and then cut the southern end of the island in half before dark.[1] But, in so doing, the landing force would suffer 1111 casualties—mostly in that maneuver element.[2] That's a full fourth of its original strength and clearly indicative of unexpected adversity.

A few months later, 9,000 Marines were to come ashore within the first 45 minutes at Iwo Jima and suffer 2400 casualties during the first day.[3] A massive bombardment one hour after the landing would cause much of that bloodshed. But, so too would machinegun fire from every direction as the maneuver element approached the southern airfield.[4]

Ambushes had occurred at both locations. But, their particulars varied. The Peleliu invasion was initially contested by a deadly sheaf of mortar shells between reef and shore.[5] Then, instead of seaward directed automatic-weapons fire, the debarking troops had encountered it from both flanks. They were forced to run through an interlocking stream of machinegun bullets at the waterline. And most of their armored vehicles were also hit by enfilading antitank fire. (See Figure 5.1.)

On both landings, all elements of the invasion force had been visible to carefully embedded high-ground cannons. Through virtual immunity to U.S. naval gunfire, those big guns had served a dual purpose. They'd provided an irrepressible big picture to the military representatives of a top-down society. That image seemed more important than any other threat.

This chapter has two goals: (1) to establish the details of the Peleliu beach ambush; and (2) to see why an infantry assault had been necessary to silence the most deadly emplacement. Until the specifics of this assault are better practiced, America should try no more amphibious landings.

U.S. Foes No Longer Depend upon 2GW Methods

As mentioned in Chapter 2, the 3GW practitioner's main tactic on defense is ambush. America's first exposure to this unfortunate reality came when U.S. Marines tried to cross the wheatfield at the front of Belleau Wood in the daytime during World War I (WWI). After establishing parameters for the USMC's current doctrine (MW), the WWI Germans had definitely qualified as a 3GW adversary. So would the German-advised IJA at Peleliu. That's why the Peleliu landing force got so badly ambushed, as will any future intrusion on Eastern territory.

What Other Nations Have Come to Understand

2GW requires strict headquarters control over almost every

aspect of combat. That's so it can supposedly apply maximum firepower. Because overwhelming force is the goal, those in charge frequently lose sight of opposition tactics. Nor do they pay any attention to the microterrain.

For an Oriental, this built-in short-sightedness of a 2GW adversary creates the chance to turn unusual circumstances into traps. Depending on the tactical mismatches and kill zones available, the bottom-up thinking Asian then makes the Western invaders' first day ashore miserable. It had occurred at Tarawa with machineguns in an offshore hulk. And it was about to happen at Iwo Jima with a shoreline emulsion and cinder terrace. Only left to uncover is what would catch the Peleliu Leathernecks off guard.

Beauty Is in the Eyes of the Beholder

The average U.S. commander becomes convinced his primary role in life is to further the steamroller effect. This, in itself, provides his Asian counterpart with an opportunity. During the invasion of any island, the Western leader stakes his unit's entire future on whether the preparatory bombs and shells will do any good. No matter what happens at the beach, he then quickly tries to cross the island and/or seize its most strategically important feature (normally an airfield). To do so in short order, he must require all subordinate elements to remain mostly upright and moving forward.

This inherent disregard for the hot lead about to fill the air is what first jeopardizes the American commander's momentum. A severe setback had occurred at Biak and so would another at Iwo Jima. Peleliu's invaders were also to meet with debilitating adversity. Until the U.S. Military moves beyond a continually "higher-tech" version of 2GW, it will continue to take excess losses during the first day of any invasion.

The Dynamics of Ambush

Successful ambushing requires three things: (1) a totally surprised victim; (2) unobstructed terrain; and (3) sufficient killing power. For a fully choreographed 2GW invasion force, the first prerequisite is automatically satisfied. Though no rifleman gets overly surprised by the carnage, his top-echelon leaders always seem to be. And while effective countermeasures aren't hard to ascertain at the front, orders must be violated to perform one.

Before the Peleliu invasion, only enlisted grunts on the U.S. planning staff could have spotted the terrain dangers at the island's south end. Three factors had contributed: (1) a hard and level surface; (2) little topsoil in which to dig; and (3) enough tiny bushes to give the intruder a false sense of security. No self-respecting ground pounder would enter such a place standing up. Scrub brush does little to impedes grazing machinegun fire. And, in the rush to cross the island, a young American leader might mistake concealment for cover.

Specifics of Peleliu's Beach Ambush

The U.S. assault troops had been badly hurt by all the flanking fire at the waterline. They would later describe it as "hell on earth."[6] (See Figure 5.1.) According to 1st Division's afteraction report, 26 LVTs had been destroyed the first day. But "unofficial estimates by assault unit commanders bring the total knocked out [at least temporarily] in excess of 60." Also disabled during the landing were 15-18 American tanks. By sundown, only five remained operational.[7]

Most devastating had been the enemy's anti-armor cannon at the Point. Its removal became the responsibility of 1st Marines' already legendary commander—Col. Lewis "Chesty" Puller. Capt. Hunt of K/3/1 was assigned the job.

> One spit of land on the extreme left flank of White Beach 1 was particularly vicious. It contained two anti-boat guns, several machine-gun positions, and entrenchments that were pouring fire into the flank of the assault force. K Company, 3rd Battalion, 1st Marine Regiment was given the mission of knocking them out.[8]
>
> — *Warfare History Network,* 23 December 2018

Why the possibility of such a gun hadn't been foreseen and then eliminated through preparatory bombardment is not clear. Its housing was so cleverly constructed as to remain invisible from overhead or offshore.

Figure 5.1: Damaged LVT(A) Provides Beachfront Cover
(Source: January 2007 photograph at the Nat, Museum of the Marine Corps, Quantico, VA, of the cover artwork to a 1st Marine Division cruise book from WWII.)

> [A] pitted spit of land that jutted out from the island and was well fortified. "The point, rising thirty feet above the water's edge," Captain George P. Hunt described, "was of solid, jagged coral, a rocky mass of sharp pinnacles, deep crevasses, and tremendous boulders. Pillboxes, reinforced with steel and concrete, had been dug or blasted in the base of the perpendicular drop to the beach. Others, with coral and concrete piled six feet on top were constructed above, and spider holes were blasted around them for protecting infantry."[9]
>
> — *Warfare History Network,* 23 December 2018

Which maneuver would be needed to overcome such a threat is now the issue. It had probably been more "situationally correct" than established procedure.

> Moving grimly ahead, they [the K Company Marines] pressed assaults upon the many defensive emplacements. Embrasures in the pillboxes and casements were blanketed with small-arms fire and smoke, then attacked with demolitions and rifle grenades. . . . A rifle grenade [finally] hit the gun muzzle itself (of the Point's primary artillery piece], and ricocheted into the casement, setting off explosions and flames. Japanese defenders ran out the rear of the blockhouse, their clothing on fire and ammunition exploding in their belts.[10]
>
> — *Bloody Beaches,* by Gayle
> Marine Corps Historical Center, 1996

1st Marine Regiment's reserve—1/1—would come ashore to try to assist its sister battalion. Then, near sundown on D-Day, an LVT resupplied K/3/1 with much needed ammunition and water. But two platoons of that company, at the extreme left of the landing zone, would be cut off from the rest of the regiment for over 30 hours. During that period, it would undergo a number of attacks from mostly infiltrators.[11] (See Map 5.1.)

> The Japanese had always used infiltration as a means to counteract massed American firepower. Peleliu, however, marked the first time that they used infiltration and raiding units as part of their defensive strategy.[12]
>
> — *Warfare History Network,* 23 December 2018

What Had Concurrently Happened with 2/1

After also being ambushed from the side at the waterline, a small part of 2/1 was briefly held up by the same inland rise that had posed such a problem for 3/1. (Look again at Map 5.1.)

> 15 to 18 September 1944. When the advance of his company was halted by fire from a hostile pillbox on the morning of 15 September, Sergeant Gerycz [F/2/1] boldly rushed forward and charged the enemy position with grenades and sub-machine gun fire which killed twelve Japanese emplaced in the position. On the following day, when his company's attack was held up, he courageously attacked a

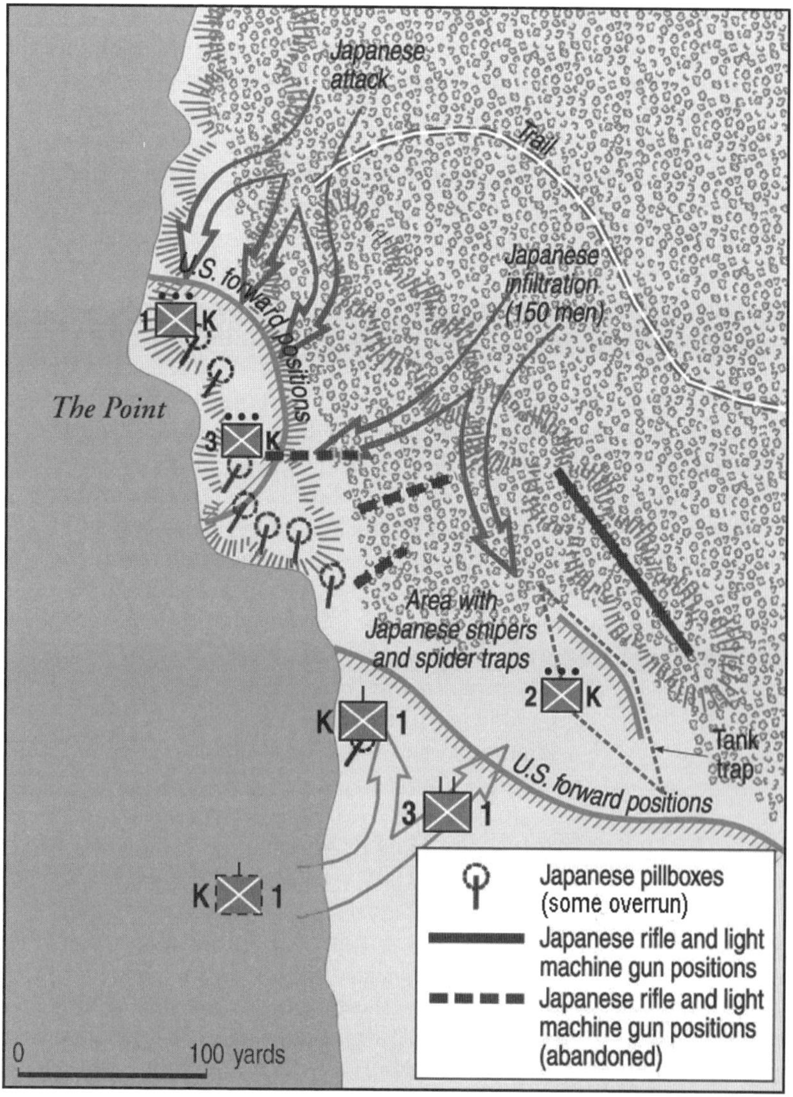

Map 5.1: Multi-Directional Strongpoint Cluster at the Point
(Source: Courtesy of Meridian Mapping, from "The Battle of Biak: 'A Terrifying Glimpse . . . ,' by D.A. Johnson, Warfare Hist. Network, © Meridian Mapping, n.d.)

> strong fortification with grenades and sub-machine gun fire which annihilated twenty-three of the enemy and reduced the hostile emplacement.[13]
>
> — Navy Cross citation for Sgt. Steve Gerycz

Meanwhile, the rest of 2/1 had moved around the south end of that rise. From the new location, it would be the first Marine unit to oppose a Japanese tank-infantry counterattack later that day. (See Map 5.2.)

> In contrast to the opposition encountered by 3/1, the 2nd Battalion . . . found relatively less resistance. Upon hitting White Beach 2, the assault companies of 2/1 spread out and drove inland. . . .
>
> Advancing inland against resistance described as moderate, the Marines . . . pushed on through the heavy woods and swamps, bypassing well-organized enemy strongholds or eliminating them with flamethrowers and demolitions, until they reached the O-1 phase line—about 350 yards inland--by 0945 [on 15 September]. Here, in the wooded area facing the airfield, the battalion made a firm contact with elements of the 5th Marines, thus securing the right flank of the regiment [1st Marines], and held up, awaiting orders to proceed to the next phase line. The 2nd Battalion remained here, however, until the following morning because of the precarious situation on the left.[14]
>
> — "Western Pacific Operations,"
> by Garand and Strobridge,
> Historical Branch, G-3 Division, HQMC, 1971

How the Other Regiments Had Fared

While juxtaposed antitank guns pummeled every boat and vehicle to come ashore, a stream of enfilading bullets greeted the emerging troops. This sideways-directed fire came from five spits of land that stuck out into the ocean at regular intervals all the way down the beach. Its curtain of flying lead had occurred despite a Japanese map (possibly captured with the 31st Army files on Saipan) that clearly showed two spits and their crisscrossing fire.[15] (Look more closely at Map 4.3.) While a prevailing wind can theoretically create such an unusual beachfront, invasion planners had failed to account for its deadly potential.

K/3/7 came in closest to the southern antitank gun. That gun had not been as big a threat as its northern sister. Still, all battalions of the 7th Marine Regiment landed in a column, with 3/7 at its front.

> On the division's right flank, Orange 3, Major Edward H. Hurst's 3/7 had to cross directly in front of a commanding defensive fortification flanking the beach as had Marines in the flanking position on the Point. Fortunately, it was not as close as the Point position, and did not inflict such heavy damage. Nevertheless, its enfilading fire, together with some natural obstructions on the beach caused Company K, 3/7, to land [to the] left of its planned landing beach. . . . Major Hurst necessarily spent some time regrouping his separated battalion. . . . His eastward advance then resumed. . . .
>
> As Hurst's leading battalion [finally] advanced, it was to be followed in trace by Colonel John J. Gormley's 1/7. Gormley's unit was to tie into Hurst's right flank, and reorient southeast and south as that area was uncovered.[16]
>
> — *Bloody Beaches,* by Gayle,
> Marine Corps History Center, 1996

After making it to shore through all the enfilade fire, 5th and 7th Marine Regiments had to reorganize in some tank ditches beyond the beach. Eventually, the 5th Marines moved out on line to try to reach the D-Day (O-1) phase line. After its left-hand unit made contact with 2/1, the combined formation started to pivot northward across the airfield clearing. Being totally exposed to foe artillery had its cost. Then, a Japanese tank-infantry column came charging out from behind the Umurbrogol's southeastern corner.[17] (Look again at Map 5.2.) Meanwhile, two battalions of the 7th Marines had been pushing eastward against the backside of Peleliu's opposite-shore defense line. And 2/7 remained in division reserve.[18]

1st Marines' Attempt to Help K/3/1 at the Point

Shortly after the initial landing, 1/1—the reserve battalion for

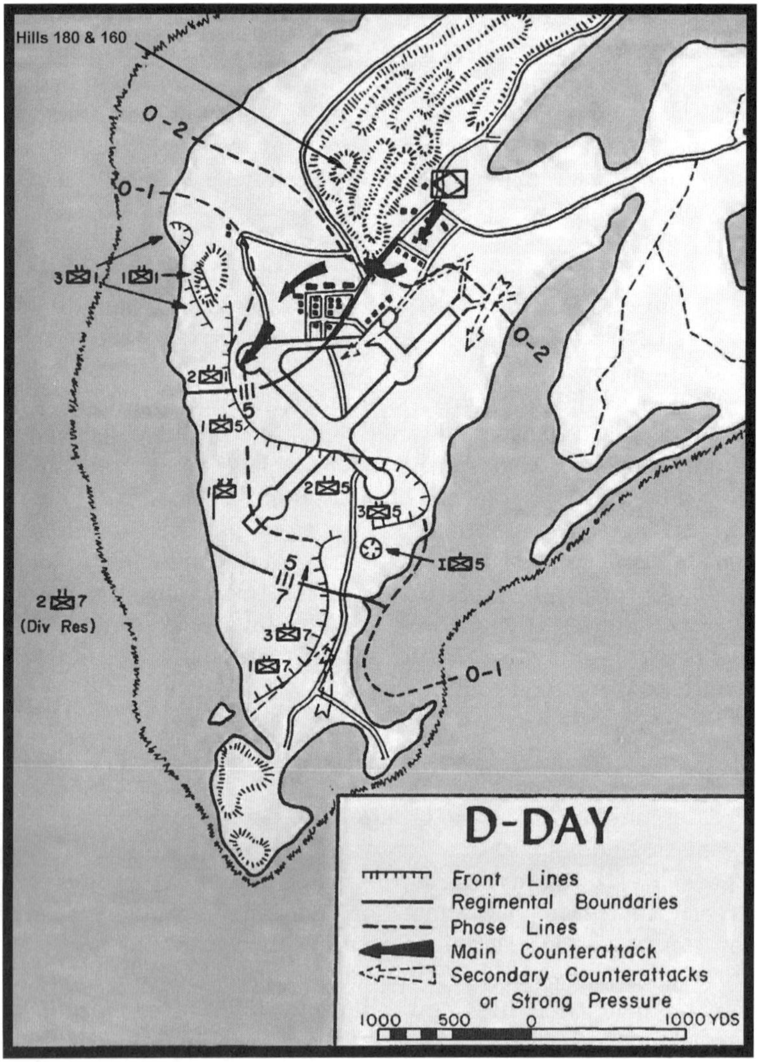

Map 5.2: Approximate Location of Marine "Lines" at D-Day's End
(Source: "Western Pacific Operations," by Garand and Strobridge, Historical Branch, G-3 Division, HQMC, 1971, p. 111.)

1st Marines—came ashore to plug the gap created by K/3/1 heading north toward the Point.[19] Part of 1/1 would be similarly ambushed (partially by snipers) in the tank ditch to the front of that same tiny rise.

> At 1300 [of D-Day], . . . Davis [of 1/1] was ordered to fill the breach [in 3/1]. The job fell to A Company, which took up position in a Japanese antitank ditch in front of the coral ridge. However, . . . "Snipers and automatic weapons sited down this ditch hit 'A' Company with damaging enfilade fire from their left flank while they were facing east toward the [coral] ridge." Davis next committed B Company to close the gap, but it was stopped cold by heavy enemy fire from the ridge.
>
> The major finally threw in C Company, his last fresh infantry unit, against the coral ridge. It managed to capture a section of the high ground.[20]
>
> — *Naval History Magazine,* October 2019

Individual squads had made the advance possible. One squad leader tried to outflank his target. So, while Chesty's boys liked to close rapidly with an adversary, they weren't opposed to a little envelopment.

> 15 September 1944. . . . [W]ith his platoon pinned down by cross-fire from two machine guns during a particularly hazardous advance through deep jungle terrain, Corporal Goff [B/1/1] boldly faced the withering barrage in a determined effort to outflank the Japanese opposing troops and, skillfully disposing his men for maximum effectiveness, fearlessly led them in a determined attack. Aware that the fire of his squad was insufficient to neutralize the heavily fortified emplacement, he pressed forward alone and, armed only with grenades and a submachine gun, succeeded in silencing the hostile weapon and annihilating the crew.[21]
>
> — Navy Cross citation for Cpl. Herbert B. Goff, Jr.

The next morning, some of 1/1 moved northward to help King Company of 3rd Battalion. After fighting through an unusual flatland complex, Company B finally made contact with the isolated Marines and returned their missing platoon. Both companies then faced increasing enemy pressure. The enemy wanted the Point back.

> 1/1 moved its Company B to establish contact with Hunt to help hang onto the bitterly contested [Point] positions.

> Hunt also regained the survivors of the [K Company] platoon which had been pinned at the beach fight through[out] D-Day. . . . That [the second] night, the Japanese organized another heavier—two companies—counterattack directed at the Marines at the Point.[22]
>
> — *Bloody Beaches,* by Gayle
> Marine Corps Historical Center, 1996

The Flatland Complex Was Part of a Composite System

A heavily fortified swath of real estate stretched all the way from the Point to the base of the Umurbrogol. Most of it lay along a low finger that projected out from that mountain's base. Besides being part of a defensive belt, this finger functioned as a reinforcement corridor. The machinegun nests along its southern edge also posed quite a threat to any airfield clearing intruder. But, what lay in the flatland between finger and Point was not what a Western soldier would expect.

> The area through which "B" and "C" Companies worked [was] rutted with anti-tank ditches and a network of camouflaged pillboxes connected by trenches on the low and flat ground. The [nearby] coral ridge [Umurbrogol finger] was honeycombed with rifle pits and machine gun nests. In particular, a pillbox with a pair of twin-mounted .80-caliber machine guns gave us a great deal of difficulty.[23]
>
> — 1/1 Afteraction Report

But, the swath's short history is what first signaled something out of the ordinary. At its southern end (in the ditch in front of the original rise), two U.S. units had been separately ambushed. Largely responsible had been an automatic cannon firing down its full length. In other words, there had be a distant gun perfectly aligned with its long axis.

> The battalion command group [for 1/1] , advancing behind the assault units, passed dead B Company Marines and a deep antitank ditch littered with the dead of A and K companies. According to 1/1's report, "It was apparent that

the Japs had sited anti-tank weapons from the left flank and mortars from the ridges to the front to cover this ditch well."[24]

— *Naval History Magazine,* October 2019

Part of 1/1 would remain atop the tiny coastal rise. Not far to the east lay three things: (1) the tip of a low Umurbrogol finger; (2) a collocated blockhouse; and (3) a road leading up to that same spot from the airfield service area.

Ongoing Events at the Point

The Japanese 47mm cannon that had done most of the shoreline damage was not of advanced design or size, just large enough to penetrate American tanks and landing craft from the side. Nor was it in a beachfront location. Because it faced southward across open water from the base of a coral promontory, its activities were not visible to any ship-borne naval-gunfire spotter. Over an hour after the initial American landings, it still bothered the 1st Marine's reserve battalion as it came ashore.[25] The only way to silence that gun was an infantry assault from the landward side.

For a modern-day American landing force to eliminate such a threat right away would take a reconnaissance swimmer with a piece of equipment that has yet to be invented—a fully submersible, shoulder-fired, and laser-guided missile. Through more emphasis on microterrain than technology, that lone gun at the Point in 1944 had become the key to Peleliu's survival.

Well enough hidden among the inland outcroppings, other 40mm and even 20mm guns with armor-piercing shells could have also ambushed any U.S. maneuver element from a distant flank. In fact, defending strategically important ground from elsewhere was about to take many forms on Peleliu. It was something a Western army never attempted with a machinegun, and thus did not expect from any opposition small arms.

The Promontory Gun Site Posed Many Challenges

Aerial photographs of the Point hadn't permitted naval gunfire to remove the offending emplacement. An afteraction picture of its rear entrance has the following caption: "The bunker was not scathed by the pre-landing bombardment."[26] According to Lt.Gen. Lewis "Chesty" Puller's biographer, preliminary reconnaissance of the island had revealed no defensive works to the right or left of where the 1st Marines were to land. But, subsequent photographs did show both promontories to be fortified.[27] U.S. planners simply hadn't expected sideways anti-armor fire. The Point's main gun had been "hidden in plain sight," either through camouflage or its proximity to other emplacements (some possibly dummies).

At any time, a 14- or 16-inch U.S. shell could have solved the problem. But, the precise location of that 47mm gun was only evident from nearby. That's what kept the assault troops from safely adjusting big shells onto it. While Chesty may have been suspicious of that promontory, the maneuver with which K/3/1 planned to occupy it still followed his well-known preference—going straight at any opposition.

> Aerial photographs of the objective "showed anti-boat obstacles on the coral reef in front of the beach, entrenchments on the beach itself, and two pillboxes," Hunt recalled. After carefully studying them, he decided that his company would "perform a gate-like maneuver by pivoting 90 degrees to the north with our left flank anchored on the beach."
>
> Hunt assigned the 3rd Platoon to clear the point, while the 2nd Platoon assaulted the right half of the company objective. The 1st Platoon was to follow in trace of the 3rd Platoon.[28]
>
> — *Warfare History Network,* 23 December 2018

Though seldom acknowledged as such, Col. Puller was actually one of the Corps better tacticians. He had personally seen a lot of action and greatly valued the junior grunt's contribution. His "direct approach" to enemy contact had its advantages. The Point gun needed be taken out immediately. A pro-active combatant is at less risk than an undecided one. And a group of charging "devil dogs" tends to disconcert any quarry expecting a more deliberate attack (like one preceded by bombardment). Having saved the airfield on Guadalcanal from a massive Japanese onslaught in 1942, Chesty further knew how much damage a defensive machinegun can do.

The only sure way to dodge its deadly stream is to get beneath it. And, going straight at the average defensive position works just fine as long as everyone crawls.

Some of Peleliu's Terrain Had Been Altered

The Point's flatland ditches hadn't occurred naturally. And the tiny coastal ridgeline to their south may have been tampered with. K/3/1 had first encountered that berm-shaped terrain feature. To its members, the five-meter elevation on period maps seemed higher. Had the tiny rise been augmented with the former contents of the ditch to its front? Then, covered trenches could have connected its bunkers. And from a ridgeline extension, a Japanese gun could have swept the ditch bottom clean.

> [M]ost of 3/1 ran into a coral ridge about thirty feet high and roughly parallel to and about one hundred yards inland from the beach. . . . The ridge was honeycombed with defensive works. Fronting the seaward side of the rise was an antitank ditch guarded by another 47mm [or lesser caliber] gun . . . on the north end [of that same ditch].[29]
>
> — *Chesty,* by Hoffman

K/3/1 Had No Time to Waste

The third platoon of Company K had avoided delay by turning north along the beach. With the first platoon in reserve, the third then drove through a number of Japanese riflemen in open foxholes.[30] IJA pillboxes of this period were surrounded by squad strongpoints. (See Figure 5.2.) And each squad's *Nambu* was protected by a ring of riflemen.

Before closing with the final objective, that "reinforced" American platoon must have knocked out the large Japanese gun that had been giving its sister platoon such a hard time.

> 3rd Platoon . . . was pinned down about 50 yards from the point. During its hard-fought assault, the platoon had knocked out one 40mm gun, two heavy machine guns, and numerous light machine guns. . . .

> The platoon sergeant, John Koval, ". . . continued pressing the attack toward a Japanese anti-boat gun.[31]
>
> — *Warfare History Network,* 23 December 2018

> 15 September 1944. . . . Platoon Sergeant Koval . . . despite a wound sustained while leading an assault against enemy pillboxes and infantrymen entrenched in spider holes along the beach, tenaciously continued pressing the attack toward a Japanese anti-boat gun emplacement which was inflicting heavy damage on our landing craft. Although wounded a second time and in a dying condition, he courageously directed the final assault and was responsible in a large measure for the destruction of the enemy gun.[32]
>
> — Navy Cross citation for Sgt. John Koval

Dirt-Covered "Octopus Pots"?

Ahead of the bunker belts at Bairoko, the Marine Raiders had faced riflemen in trees and "*Nambu* machinegun[ners]s in outpost foxholes."[33] At the northern promontory on Peleliu, they ran into both varieties of adversary in "spider holes" and "spider trenches."[34] (Look back at Map 5.1.)

> The spider trenches were manned by riflemen and light machine gunners.[35]
>
> — 3/1 Afteraction Report

The generally accepted definition of a spider anything is one with a lid. That way, it can function as either a forward outpost or ambush. The ambush option is the most interesting. As long as the quarry looks away when the position opens fire, he can later stand atop that position without realizing it. On Peleliu, such emplacements existed in swampy ground between Point and tiny coastal berm. (Look back at Map 5.1.) "Octopus pot" is what the Japanese called each fighting hole in a squad strongpoint array (light machinegun pit encircled by rifleman holes).[36] There would be many octopus pots on Iwo Jima and Okinawa.

> [A Japanese] POW [prisoner of war] can recall . . . specific . . . recommendations made in the [32nd Army] Battle Lesson(s):

> . . . that they increase the distance between the so-called "octopus pots" *(Kakutsubo)* foxholes in preparing defensive positions in front of caves and similar entrenchments.[37]
>
> — Col. Hiromichi Yahara, survivor of Okinawa

On Peleliu, the 7th Marine Regiment would also take fire from obscure dugouts in the mangrove swamps to the east of the airfield. Because that fire didn't always come from the front, it had the effect of an ambush. Nowhere in the discussions of Peleliu are snipers in beachfront palm trees mentioned (like on Guadalcanal). Only the spider hole variety had been evident in the Palaus. Biak had also produced snipers in ground level "crevices and recesses."[38]

Final Seizure of the Promontory Gun

There had been Japanese machinegun nests just north of the Point's antitank gun. They became the job of 1st platoon, K Company. (Look more closely at Map 5.1.) Cpl. Hahn was likely a member of that platoon.

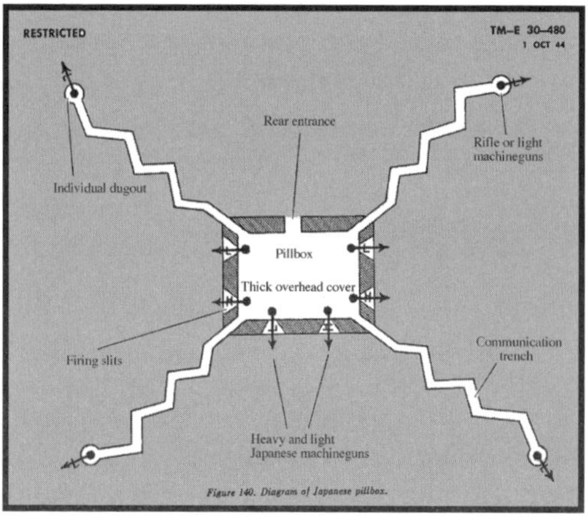

Figure 5.2: Pillbox at the Center of Squad Strongpoints

(Source: "TM-E 30-480" [1944], p. 160.)

> [O]n 15 and 16 September 1944. Observing four of the enemy setting up a heavy machine gun in a position which seriously threatened his squad, Corporal Hahn [K/3/1] crawled forward and, with unerring aim, tossed hand grenades into the hostile position, killing three of the Japanese and forcing the other to flee. Having captured the gun, he continued to lead his men in a furious running assault against hostile positions defending the north end of the beach, inflicting heavy casualties on the hostile troops and winning new positions for the friendly forces.[39]
>
> — Navy Cross citation for Cpl. Henry W. Hahn

In all, five reinforced concrete pillboxes beneath many feet of coral rock would eventually be captured on the northern promontory. They had housed four 25mm machine cannons and the offending 47mm antitank gun.[40] But, any tactical insights must derive from the better reported take-down of the big gun.

> From 15 to 17 September 1944. A bold and aggressive leader, Captain Hunt [K/3/1] led his men in a daring assault against the enemy who were firing from concrete pillboxes on a coral point. Knowing the great danger the seizure of the point would incur, but realizing the immediate necessity for its capture, he quickly and skillfully maneuvered his company and, with two platoons, captured the point after a fierce struggle during which five hostile concrete pillboxes, numerous coral pillboxes and lighter emplacements were destroyed and over one hundred of the enemy were killed.[41]
>
> — Navy Cross Citation for Capt. George P. Hunt

Most Likely Assault Method

The full "blind 'em, burn 'em, and blast 'em" routine was not applied to the Point's main gun bunker. That method's limitations would be fully evident from the Tarawa blockhouse scenes in "The Sands of Iwo Jima."[42] They had been well known to every member of 1st Marines. Nowhere in any description of the Point struggle is there any mention of flame, so Chesty's momentum-oriented "Jarheads" had modified the second and third steps of the procedure. They had blinded bunker occupants with bullets and white phosphorous smoke, then tried to get a tiny explosive through the gun port. For any long and flat approach, a rifle grenade was the only way to accomplish the latter. (See Figure 5.3.)

> 15 September 1944. Ordered to seize a thirty-foot high coral point of land which commanded the entire landing beach and from which the enemy were delivering devastating anti-boat and heavy machinegun fire on our personnel and landing craft on the beach, Second Lieutenant Willis [of K/3/1] led his men forward in a daring and skillful assault. During the fierce hand-to-hand conflict which reached a high pitch of intensity when he and his men had penetrated the Japanese ring of infantrymen and were assaulting the pillboxes themselves, Second Lieutenant Willis boldly crawled among the coral boulders with one other man to the flank of a concrete pillbox that was firing an anti-boat gun and, maneuvering himself to an advantageous but precarious position, hurled white phosphorous grenades at the embrasure, thereby enabling his comrade to encircle the pillbox and throw grenades into the exit.[43]
>
> —Navy Cross citation for 2ndLt. William L. Willis

> Since The Point's fires were oriented primarily towards the landing beach area, the Marines decided to assault the bastion from the . . . (east). Gathering up the remnants of the 3rd Platoon, Second Lieutenant William L. Willis and his 1st Platoon began slugging their way toward the top of the objective in the face of concentrated enemy fire. After killing off its protecting infantrymen, the Marines approached each pillbox from its blind spot to blast the occupants with grenades. Finally, the attackers broke through the maze of infantry positions and pillboxes to storm the crest of The Point, but beneath them at the water's edge could be heard the roar of the 47mm cannon that had wreaked havoc among the Marines all morning.[44]
>
> — "Western Pacific Operations,"
> by Garand and Strobridge,
> Historical Branch, G-3 Division, HQMC, 1971

The Point's main installation turned out to be a reinforced concrete casemate built into the cliff base of a coral promontory. Its

47mm antitank/anti-boat gun faced southward across open water. As such, elements of K/3/1 were forced to approach it from the east through several squad strongpoints ("spider trenches . . . manned by riflemen and light machine gunners"). Then, the gun was finally silenced by "creeping down on it from above." Lt. Willis dropped a smoke grenade just outside the embrasure, so Corporal Anderson could get close enough to launch a rifle grenade into the firing aperture. This grenade struck the gun barrel and then ricocheted inside the bunker, starting a fire.[45]

> "A squad covered the rear exit of the pillbox," he explained. "Anderson, one of my corporals, sneaked part way down the rocks about twenty yards in front of the embrasure, while I crawled to a cut in the cliff where I could heave a grenade without being shot."[46]
>
> —Lt.Willis, as quoted by *Warfare History Network*

Figure 5.3: Approaching a Bunker Across Flat Open Ground
(Source: "Follow Me," by H. Charles McBarron, at Leyte, 1944, U.S. Army Center of Military History, artphoto archives, with illustration designator, "avop05-99_1.jpg,")

So, three things had occurred: (1) Chesty's boys had planned on and used rifle grenades to handle the additional distance to the bunker aperture; (2) they had tossed white phosphorous grenades to screen some of their movement; and (3) they had not crawled the whole way to the target. While not as sophisticated as a WWI Stormtrooper assault, this procedure still qualifies as a "quick deliberate attack." Something like it had been practiced ahead of time, and its final conduct entailed some scouting ("recon pull").

Where Was the Rest of 3/1 All This Time

Along with Company K's second platoon, Company I and now Company L were still contending with the fortified berm that ran southward behind the beach. (Look back at Map 5.1.) Company I's bunker assault was similar to that at the Point. Only here, more cover during the approach had permitted hand grenades through the gun port.

> 15 . . . September 1944. When his platoon was pinned down by heavy Japanese machine-gun fire from two pillboxes, Private First Class Case [I/3/1] fearlessly exposed himself to the terrific barrage and, crawling forward to an open, more advantageous position, directed extremely effective automatic rifle fire on the embrasures of both emplacements. Then, advancing boldly in company with another Marine, he assaulted the hostile position with determined aggressiveness and, hurling grenades and engaging the enemy in hand-to-hand combat, succeeded in annihilating the eight Japanese occupants, thereby enabling his platoon to continue the attack.[47]
>
> — Navy Cross citation for PFC Victor L. Case

A lieutenant from I/3/1 then borrowed a trick from the Asians. Instead of going directly toward each berm bunker, he moved to one side and took it under fire from a flank (possibly with a bazooka).

> 15 September 1944. Although wounded by enemy shrapnel and hand grenades shortly after landing, Second Lieutenant Cladin [I/3/1] refused to be evacuated and insisted upon leading his platoon in an assault on hostile pillboxes which

> were delaying the advance of his company. Expertly maneuvering his men into position, he directed flanking fire on the Japanese emplacements and succeeded in destroying them.[48]
>
> — Navy Cross citation for 2ndLt. Lewis Cladin

Such a stratagem is immensely important. To safely destroy an enemy emplacement, one cannot head directly toward its gun port. Having come to the same conclusion, the Maoist Raiders at Bairoko had likely come up with an indirect way for each fireteam to approach the bunker in its lane—through a double-envelopment by buddy teams. That way, one pair of "Gyrenes" could attract the occupants' attention, while the other pair moved closer to the target. It was through this same doctrine-violating maneuver, that 2/4 (the former 4th Raider Battalion) would likely seize the rear face of Half Moon, permitting Okinawa's infamous Shuri Line to be finally breached in 1945.[49]

Not going directly at any particular enemy machinegun would also help the badly outnumbered Marines of 2/4 to win the Battle of Dai Do in Vietnam in 1968.[50] Instead, they moved into a "gap" between emplacements and fired down the adjoining trenchline.

Then Came the Large Point Counterattack

Recapturing the Point cannon had clearly been part of the Japanese plan to push the American invaders back out to sea. Their troops made a concerted try at that gun site on the night of 16 September.

> By mid-morning D plus 2, Hunt's survivors along with Company B 1./1 owned the Point, and could look out upon some 500 Japanese who had died trying to re-take it.[51]
>
> — *Bloody Beaches,* by Gayle,
> Marine Corps Historical Center, 1996

This enemy push had been against all 1st Marine Regiment units in the area. As such, it may have initially come from the Reserves' cavern inside the Umurbrogol. The coastal berm that had held up the 1st Marines' advance inland was thin and no more than 15 feet high, according to Map 4.1. But, within 150 yards of its northern end was the wider and somewhat taller western extension of Umurbrogol Mountain. That second more ominous feature is what the following excerpt seems to address.

> Major Davis ordered us to "dig in." His forward command post [CP] "was a little more than a hundred yards in from the water."
>
> About another 100 yards inland [a total of 200-300 yards from the beach], a heavily defended 30-foot-high long coral ridge was posing a serious problem for 2/1 and parts of 3/1, on the front line ahead of Davis's supporting battalion. Covered in foliage, it had gone unnoticed during pre-invasion reconnaissance of the island, Davis [CO of 1/1, the Regimental Reserve] recalled.[52]
>
> – *Naval History Magazine,* October 2019

Map 5.1 fails to show the tip of this squat Umurbrogol finger, but Maps 4.1 and 4.2 confirm it was there. From a spot within 300 yards of Peleliu's shoreline, it ran eastward to the base of the massif. Map 5.1 does reveal the end of a trail where the big Point counterattack was first spotted. That trail led back to the finger tip. So did a road from the airfield. But with all the aerial illumination around after dark on 16 September, a Japanese battalion couldn't have made it all the way up that road without being destroyed by U.S. naval gunfire. Nor could it have moved 600 yards cross country from the base of Umurbrogl Mountain.

At the end of that same finger was a water well and some ruins. Within the well may have been the exit to a long sub-surface passageway. Before the landing, a huge Japanese blockhouse had been constructed at the same spot with a dozen supporting bunkers, all connected to the parent structure by tunnel.[53] Such an elaborate below-ground network suggests a connecting conduit to IJA headquarters.

One arrow on Map 4.4 shows the Point counterattack coming from the base of the Umurbrogol. Without a tunnel, how could that large an attack force have gotten unnoticed to the end of a largely barren finger?

6 Peleliu's Developing Ambush Pattern

- What had been the tactical trends so far on Peleliu?
- Did long-range fire offer any advantages?

37mm Model 94 antitank gun emplacement from 1944.

(Source: "Japanese Defenses and Fortifications . . .," by J. David Rogers and Karl F. Hasselmann [Univ. of Missouri at Rolla, n.d.], p. 27.)

Every Asian Defense Plan May Follow a Different Ruse

East Asia is where the modern art of military deception was first introduced by Sun Tzu. If a 3GW-like defense is largely through ambush, then its variants on Peleliu all derived from a new form of entrapment. At Bairoko, there had been tiny bursts of machinegun fire in front of an MLR. On Biak, a mechanized thrust toward the airfields had been met with a *Haichi Shiki*. But, the Peleliu surprise was to be as demoralizing as it was deadly. Like a bad dream, it kept "morphing" into a similar—and seemingly inescapable—predicament.

There had been a pattern of sorts between the beach debacle and twin ambush at the same tank ditch (for A and K Companies).[1] In both places, heavy enfilade fire had occurred from somewhere else. That had likely been the result of a mutual strategy. If that strategy were totally different from Western defensive doctrine, it might work repeatedly on Peleliu. The standard U.S. MLR incorporates interlocking streams of machinegun bullets, but none are from a great distance away. And all anti-armor fire is directed outboard to the full extent of its "effective range."

So, the Yanks wouldn't expect their historically linear assault formation to be contested from far to one side—for a single enemy gun to engage an entire unit through a few turns of its elevation wheel. And, that same gun could be made almost immune to return fire by how it was mounted. As in the chapter image, it could shoot through a tiny recessed opening. Then, only GIs in the impact area could see its muzzle flashes or effectively return fire. This is an old urban warfare trick that would be documented for Phase Line Green in the Vietnam War. By placing scores of riflemen in recessed positions with overlapping fields of fire, an NVA commander created an uncrossable street.[2] And only the procedure's victims could tell what had happened. But in WWII, a recessed-position fire plan would not be reported until the defense of Kuribayashi's cave in Iwo Jima's narrow Gorge.[3]

That's what was behind Peleliu's recurring nightmare. But, tiny Bairoko-like ambushes would also occur near defense lines, and a Biak-like *Haichi Shiki* confront at least one U.S. road march. Peleliu was the final testing ground for all defensive tricks used at Iwo Jima and Okinawa.

The Enemy's Thought Process

The Nipponese had known the American 2GW steamroller to be frontally and surface oriented. It deployed most of its combat power on line, and then went after enemy troop concentrations. So, the soldiers of the Rising Sun planned to counter that steamroller with a 3GW surprise—initially from a flank and then from behind. In that way, they could impede every attack and then undermine its means of support. Almost by definition, a 2GW proponent would be unperturbed by indirect opposition.

As the Russians had done at Stalingrad, the Japanese would also channelize segments of the U.S. steamroller into a series of enfilade ambushes. The last of those ambushes would occur in the sheltering ditch to the immediate front of its next objective.

More About the Defense System Between Point and Tunnel

Most of the initial resistance had come from Peleliu's northern promontory. Then, the Japanese would try to take that promontory back. To help in that regard, there had been a clever defense grid between Point pillbox cluster and Umurbrogol finger tunnel to keep the intervening 250 yards from too quickly falling into U.S. hands. That grid involved the unexpected interaction between antitank ditch and machinegun bunker string. Normally the two occur back-to-back across the path of an intruder. The first makes the second less easy to overrun. But not this time. Between Point and Umurbrogal finger lay a strange patchwork of the two features. And some bunkers were shooting down the full length of a distant and non-associated ditch. In that way, they could ambush GIs staging for the assault on almost any objective. (See Figures 6.1 and 6.2.)

> These antitank ditches [like the beaches] . . . were covered by [flanking] fields of fire from pillboxes and gun casemates, located in dominating positions and all linked together in a system of mutual cover and support.[4]
>
> — "Western Pacific Operations,"
> by Garand and Strobridge,
> Historical Branch, G-3 Division, HQMC, 1971

Companies B and C of 1/1 had first noticed that flatland maze of tank ditches and bunker rows as they went to the assistance of K/3/1.[5] With little other choice, they had focused on the emplacements to their front.

> 16 September 1944. When the advance of his unit was held up by a deep zone of strong enemy emplacements consisting of a series of conccaled concrete pillboxes and rifle pits, and the *cross-firing of heavy weapons swept the front of his company,* Sergeant Monarch [1/1 squad leader] skillfully maneuvered his men into advantageous positions to deliver

> accurate covering flame-thrower and rifle fire against the Japanese strong points. Courageously and with aggressive determination, he repeatedly made his way forward in the face of intense, continuous hostile opposition and personally placed heavy explosive charges into the embrasures of the enemy emplacements, thus destroying ten of the strongholds and enabling his company to advance.[6] [Italics added.]
>
> —Navy Cross citation for Sgt. Charles R. Monarch

In effect, the Japanese had made every ditch to the front of a bunker row into a kill zone for anyone trying to attack that row. Completing the trap had been barbed wire between ditch and row, plus a preregistered mortar sheaf.[7]

> "It was apparent that the Japs had sited anti-tank weapons from the left flank . . . to cover this ditch . . . and . . . made the ground untenable."[8]
>
> — *Naval History Magazine,* October 2019

Such an arrangement constituted a tactical advance over the shrapnel-covered bunker belts of Bairoko. Now, the Japanese could empty the staging area for every American infantry assault. And a zigzag defense line became self-cleansing. (Look at all the tank-ditch fire in Figure 6.1.)

GIs assigned a narrow assault lane would be particularly vulnerable to long-range flanking fire. And, 20mm machine cannons can accurately shoot up to 1600 feet, whereas .50 caliber machineguns have an effective range of 2000 yards.[9]

There's little doubt Japanese headquarters had orchestrated all the ditch ambushes. The National Park Service investigation reported, "anti-tank ditches skillfully located so as to channel enemy tanks toward [the kill zone for] anti-tank weapons."[10] What happened on Peleliu may not have been as orderly as in Figure 6.2, but it had the same effect on the Americans.

Why Hadn't More Attention Been Paid to Enfilade Fire

2GW-oriented forces tend to follow their pre-arranged attack plan irregardless of any unforeseen adversities or opportunities. Speed-driven momentum was headquarters' apparent intent

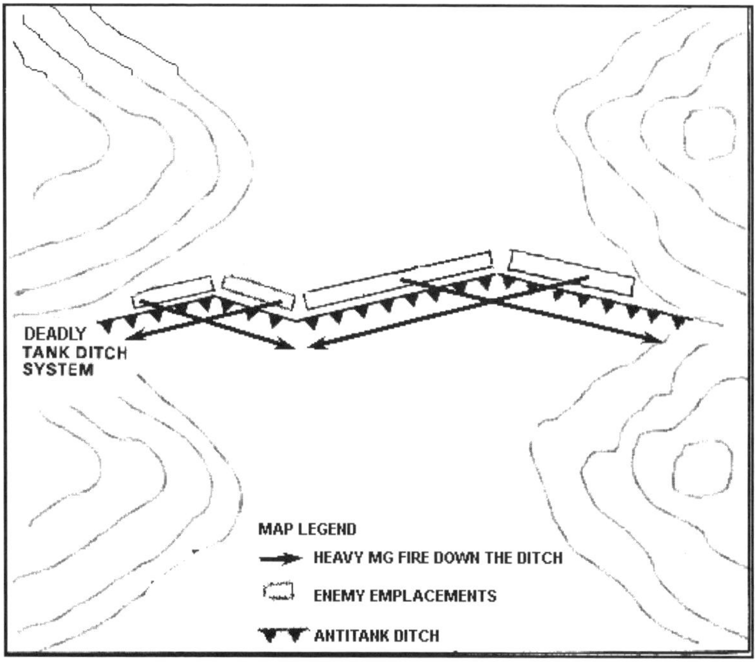

Figure 6.1: Automatic Tank Ditch Ambush
(Source: FM 5-102, "Countermobility," fig. 6-8.)

throughout the Peleliu invasion. From high overhead, all linear terrain features look about the same (with differences in width and composition difficult to establish). So, Peleliu's antitank ditches and communication trenches became indistinguishable. And, any mixture of the two with emplacement strings would have well disguised the threat.

Two Marine companies had been separately waylaid in the same tank ditch.[11] And Peleliu's defenders could have just as easily turned any straight stretch of communications trench into an ambush site. Fortunately, the Marines would have considered any forward-leading segment too risky to follow.

Neither was the airport clearing identified as a kill zone for long-range automatic weapons. Barely visible in pre-invasion Maps 4.1 and 4.2 were two slightly elevated areas on either side of that clearing: (1) a low finger extending westward from the Umurbrogol; and (2) a coral outcropping across the southern tip of the island. Both of those tiny rises had hosted machinegun nests. Between them lay the island's main runways and 2000 meters of equally level coral shelf. Upon that hard shelf lay a few inches topsoil and the occasional bush.[12] Just to the north of the airfield were the ruins of its service area. Within that area, Japanese Map 4.3 shows a cluster of strongpoints that would have minimally contained .50 caliber machineguns with a range of 1800 meters. Yet, much of the U.S. landing force still dutifully lined up across this veritable "skating rink" on its first day ashore. So doing hadn't helped its casualty count. Nor would infantry protocols that discouraged night movement and crawling.

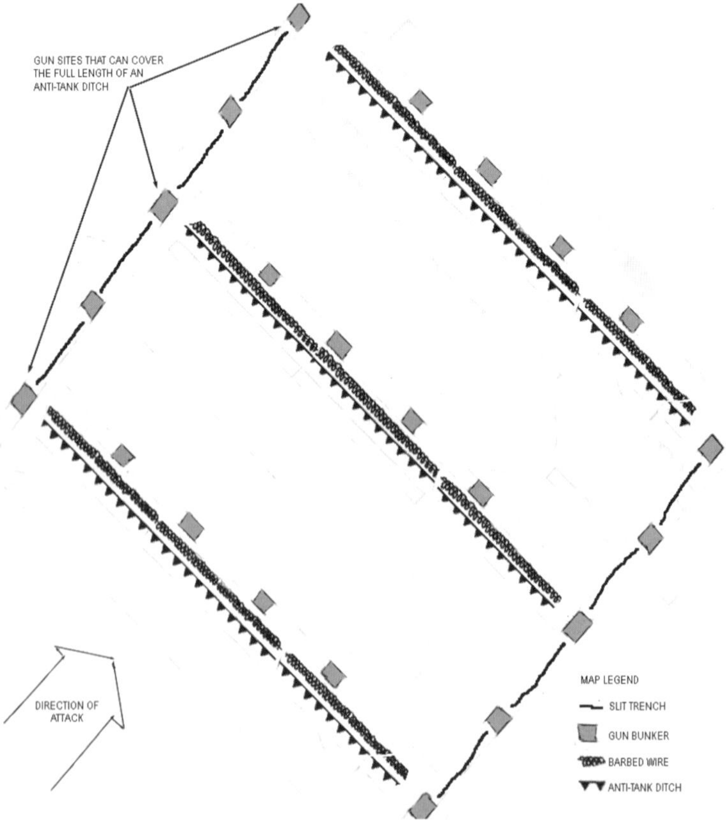

Figure 6.2: Peleliu-Like Tank Ditch Configuration
(Source: FM 5-102, "Countermobility," fig. 6-8.)

Enfilade Killing Can Also Occur at a Smaller Scale

At the most assault-resistant objective on Iwo Jima in February 1945 (the Japanese MLR's notorious Amphitheater), enemy engineers had constructed a defense matrix with marked similarities to Peleliu's tank ditch ambush. (See Figure 6.3.) Instead of distant bunkers that could shoot down the long axes of tank ditches, they built intermediate positions between successive rows of bunkers. Through tiny sideways-facing gun ports in these seemingly extraneous positions, Japanese soldiers could secretly snipe Leathernecks heading for the next row of emplacements. In all the din and confusion of heavy combat, the victims would never know what hit them.

Figure 6.3 is a drawing from a POW who had just helped to defend the Amphitheater on Iwo. It shows a low ridge, linear mound, or dirt-filled depression containing a perfectly rectangular cluster of some 60 machinegun emplacements.[13] Such a well-ordered array could not have been hewn out of solid rock; it must have been constructed above ground and then partially buried. About 20 Japanese infantry squads (two companies with 15 men and one machinegun per squad [14]) could have gradually shifted rearward through such an array. Instead of tunnels, they would have had covered trenches. So, Peleliu's "flank ambush" would bother other Americans in 1945. Hopefully, it won't come back to haunt future GIs.

A 2GW Scheme Had Come Up Against a 3GW One

The first part of the Peleliu fight had proven overly costly for the Americans. All U.S. maneuver elements had been ambushed and then mostly counterattacked through infiltration. There is no shame in this for any U.S. participant, of any rank. He had done well under conditions for which his style of war made too little provision.

U.S. forces would win out in the end at Peleliu. This victory serves as a reminder for future operations. Such battles can only be decisively won through small-unit action, and never bombardment. Failing to search below ground right away for the enemy's command and logistics hubs will lengthen their duration. And, all frontline fighters must be allowed to counter enemy tricks through adjustments to their own tactical technique.

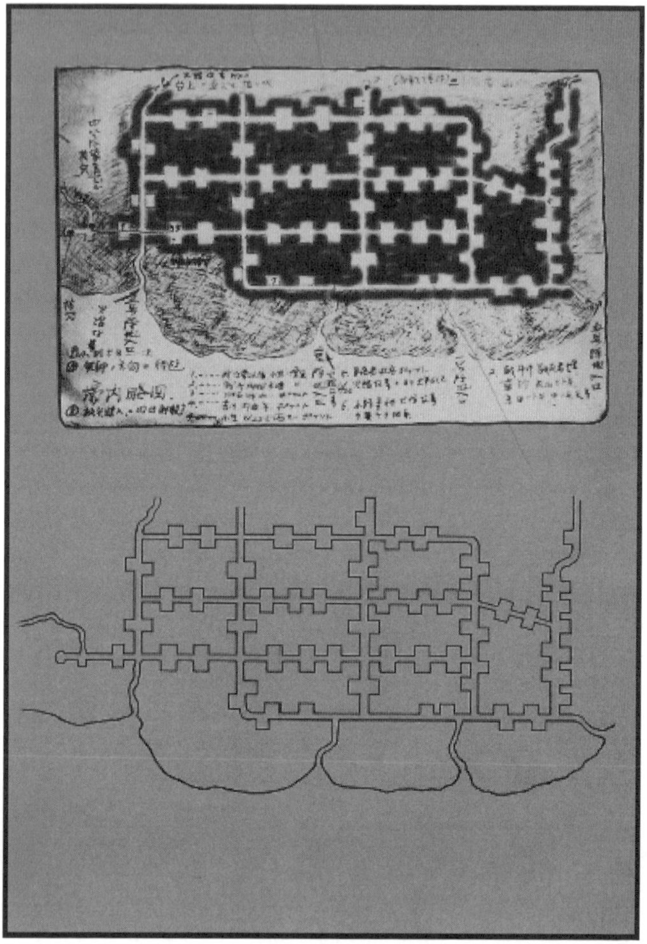

Figure 6.3: Sideways-Looking Positions Between Bunker Rows

(Source: "Japanese Defenses and Fortifications Tawara, Iwo Jima, and Okinawa 1943- 1945," by J.D. Rogers and K.F. Hasselmann, Univ. of Missouri at Rolla, n.d., p. 54.)

7 Extent of Subterranean Chambers on Peleliu

- Were many Japanese positions linked by tunnel?
- How far did the enemy travel below ground?

Some pillboxes had sub-surface extensions.

(Source: "Bloody Beaches: The Marines at Peleliu," by B.Gen. G.D. Gayle, Marines in WWII Commemorative Series. p. 47.)

Had Peleliu Been Secretly "Hardened" Since 1920?

In 1923, Marine Lt.Col. Earl Hancock Ellis died under suspicious circumstances on Koror in the Central Palaus while looking into how well Japan had followed its League of Nations mandate.

> [Japan's] title to all former German Pacific colonies north of the Equator was legitimized . . . by League of Nations Mandate in 1920. So far as the Palaus were concerned, this title remained undisputed, even after Japan withdrew from the League in 1935. . . .

> Under the terms of the [WWI] Mandate, they were specifically prohibited from fortifying any of the territory entrusted to them, but the extreme secrecy with which the Japanese immediately cloaked all their activities and the pointedness with which they attempted to exclude all foreigners soon gave rise to lively suspicions that they were doing so.[1]
>
> — *The Assault on Peleliu,* by Hough
> Historical Division, HQMC, 1950

The Japanese had ostensibly tried to extract phosphate from Peleliu, but the attempt was short-lived.[2] A phosphate refinery at the north end of the island and open-pit mine at the other end may have provided the perfect cover for defensive excavations. Map 4.1 reveals crossed spades (gravel, sand, clay, or borrow pit) near the southern tip of Walt Ridge. Bat droppings in natural caverns also produce a limited amount of phosphate. And Peleliu did have the best airstrip in this part of the western Pacific.

> Earliest wartime [WWII] intelligence indicated that the Palaus were being used mainly as a staging area and replacement center for troops, aircraft and naval units destined for the Netherlands Indies and New Guinea.[3]
>
> — *The Assault on Peleliu,* by Hough
> Historical Division, HQMC, 1950

Peleliu's Natural Cave System

Peleliu mostly consisted of coralline limestone, some of which had been forced upwards through ocean bottom pressures. That limestone had since been eroded by yearly monsoons. A number of caverns, caves and interconnecting channels had resulted from rainwater runoff, as on Biak. Many of the bigger chambers on Peleliu had been utilized for storage and other things for centuries by island dwellers.[4] After WWI, the Japanese brought in engineers and laborers to turn those naturally occurring spaces into a sophisticated defense network. Just how much the underground streamlets were enlarged, no one really knows. But, in combination with brand-new tunnels, they may have stretched quite a distance.

> By the summer of 1944, the Japanese had . . . about 11,000 . . . [troops] on Peleliu . . . augmented by Korean and Okinawan laborers. These workers had honeycombed the island with miles of underground bunkers, living quarters, and fighting positions. . . .
>
> . . . [O]n the morning of September 15, 1944, the 1st Marine Division headed for Peleliu against the fanatical Japanese, who, burrowed deeply in their subterranean fortress, were largely immune to the thousands of bombs and shells.[5]
>
> — *Warfare History Network,* 23 December 2018

> In the Umurbrogol, the Japanese Army on Peleliu had been digging "like moles" to create their defenses. Using the natural coral outcroppings that stretched over 5,000 yards from the airstrip to Hill 80 in the northern region, they constructed a "monster Swiss cheese, pocked . . . with caves and crevices." Some were tiny and could only hold a lone rifleman; others were so immense they held entire battalions.[6]
>
> — *Warfare History Network,* 4 January 2019

Immediately following the Peleliu battle, the Island Command counted over 500 caves. The Japanese Army had tried to incorporate such caves into a strongpoint matrix. But, the Japanese Navy only wanted them as bomb shelters for its coastal artillery. So, the IJA occupied all sub-surface chambers with any tactical value, improved upon their natural attributes, and then excavated others for mutual support. And the Japanese naval contingent dug most of its own underground facilities. To help in this regard, it had the 214th Naval Construction Battalion—a tunneling outfit *(Suidotai).*[7]

Full Extent of the Naval Excavations

According to the original battle chronicle, "[M]ost of the Navy caves were located near the northern end of the . . . peninsula, though a few had been constructed adjacent to Army installations for liaison purposes."[8] Those northern guns had also protected Ngesebus. But their shells couldn't reach the airport area because of the hill-mass in the way. So, additional naval weaponry must have been placed at the airport end of the Umurbrogol. In fact, a Park Service study and the *Warfare History Network* both confirm their presence on the front face of that mountain. Another study collocates the Navy Command Cave with IJA Headquarters (after it was moved to the thin ridge between Death Valley and the China Wall). The *Warfare History Network* admits, "Nakagawa had . . . anticipated that the Marines would land in the southwest section of Peleliu . . . [and] placed . . . huge naval guns [there]."[9] Several 200mm anti-boat naval cannons were being installed at the time of the invasion. And, during the landing, large-caliber shells did come from the sharply rising slope behind the airfield.[10]

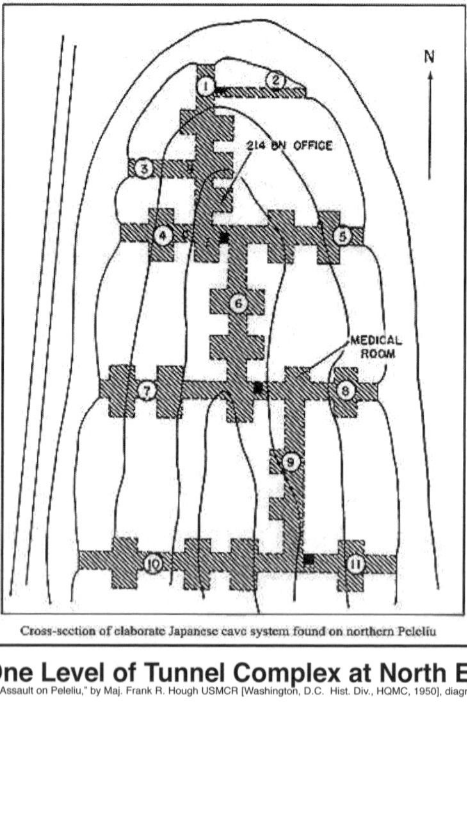

Cross-section of elaborate Japanese cave system found on northern Peleliu

Figure 7.1: One Level of Tunnel Complex at North End of Island
(Source: "The Assault on Peleliu," by Maj. Frank R. Hough USMCR [Washington, D.C. Hist. Div., HQMC, 1950], diagram, p. 195.)

> [H]ere [beyond the beach], the companies [of all three Marine regiments] were obliged to remain, owing partly to . . . the murderous artillery and mortar fire from the [initial] high ground to the north which now swept the open airfield to their front.[11]
>
> — *The Assault on Peleliu,* by Hough
> Historical Division, HQMC, 1950

Within one battle chronicle is the photograph of a gun with a muzzle of at least 155mm in diameter. It's caption reads: "Situated in a cave overlooking the airfield is this heavy caliber Japanese anti-boat gun . . . [with] a field of fire which included the invasion beaches and the airfield."[12] The Presidential Unit Citation for 1st Marine Division confirms the landing force had been heavily shelled.[13] So, other Japanese naval construction battalions must have been tunneling at Peleliu's southern end and possibly beneath its flatland shelf. Also present on the island had been the 204th, and elements of the 43rd and 235th.[14]

All new naval passageways were of ample size and free of stalactites and stalagmites. They had cross-tunnels and entrances along either side for the protection of their occupants. Some Japanese troops would occupy a "1000-man cave" deep in Amiangal Mountain until February 1945.[15] But, what tourists now visit of similar name is actually a series of passageways and rooms, to include a small hospital. So, that one big cave had actually been a multi-level underground complex. Its tunnel system was so elaborate that GIs first thought it to be a phosphate mine. A phosphate refinery had existed at the north end of the island,[16] but the only nearby phosphate diggings were by open-pit method at the west end of the Amiangal.[17]

Never designed to withstand infantry assault, those underground naval emplacements could still be defended. Much of their security would come from the abundance of passageways and bays. Figure 7.1 reveals only the lower level of an excavation at the same location as the 1000-man cave. The 5th Marines would still have trouble taking the Amiangal ridges. But, because the naval positions had not been mutually supporting, they were more easily captured than those atop the Umurbrogol.[18]

Of the many underground chamber entrances eventually photographed on Peleliu, few show any evidence of the stalactites or

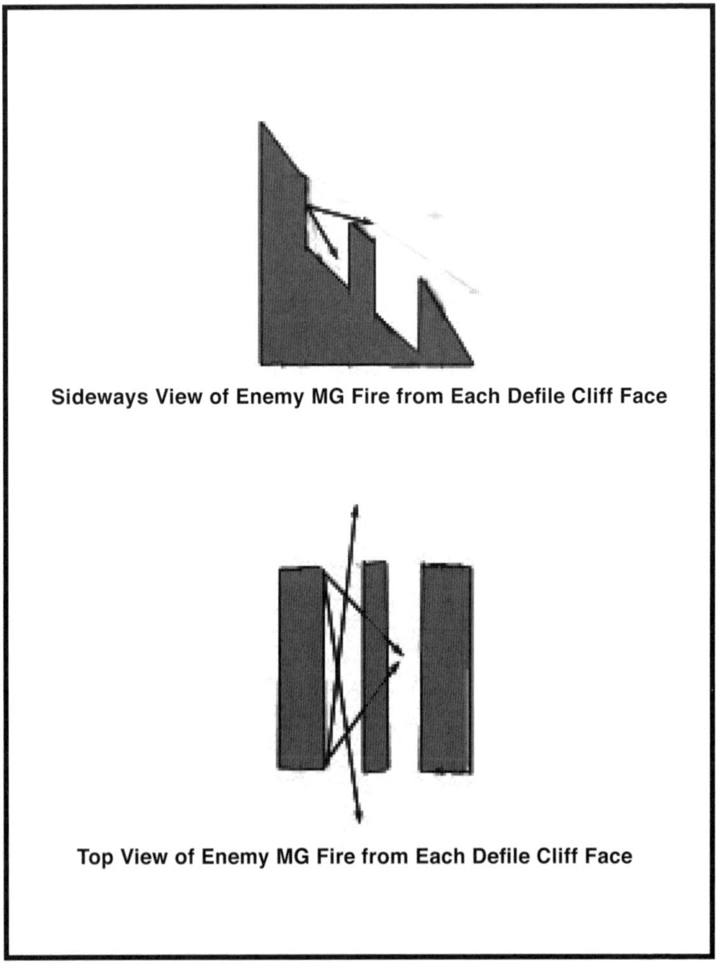

Figure 7.2: How Parallel Crevasses Were Protected

stalagmites prevalent on Biak. Most Peleliu openings were nearly round with a level floor, as if improved by Japanese naval construction crews.

Unusual Terrain at One Corner of the Umurbrogol

The southern part of Peleliu Island's central hill-mass was a single mountain with several peaks. Host to a huge underground defense complex, that mountain became known as the Umurbrogol. However, one corner contained something not often found in nature—parallel bands of hillside with deep crevasses between them.

> The mountain was a series of broken coral ridges, narrow valleys, and rugged peaks, some rising to 550 [actually 300] feet, pocked with caves and crevices. Thick jungle scrub cloaked the slopes. . . . "[W]hen we denuded it . . . , we found these funny shaped ridges [along one flank] that were as steep as a roof of a house. And instead of one ridge . . . , there might be three or four parallel ridges with deep ravines in between."[19]
>
> — *Naval History Magazine,* October 2019

Along the Umurbrogol's northwestern crest ran a high cliff. Because the narrow bands laterally traversed a steep slope, they are normally assumed to be near that mountain's summit. By fortifying them, Japanese engineers had created a virtually impenetrable defense belt segment. To get past it, U.S. troops would have to climb down into a series of narrow defiles, each swept by automatic weapons fire from above and both ends. (See Figure 7.2.) That's because small cavities had been carved into the each successive cliff face. All had rear entrances with eventual access to the central tunnel system. (See Map 7.1.)

> The ultimate desideratum [necessity] was another entrance, or entrances, on the opposite side of the ridge or out of view of the attackers around a shoulder, through which the defenders could duck out for a safe cigarette or two during such time as the Marines were earnestly blasting and burning their empty refuge. In that region of jumbled razorback ridges, construction of multiple entrances proved practicable in a surprising number of cases.[20]
>
> — *The Assault on Peleliu,* by Hough
> History Division, HQMC, 1950

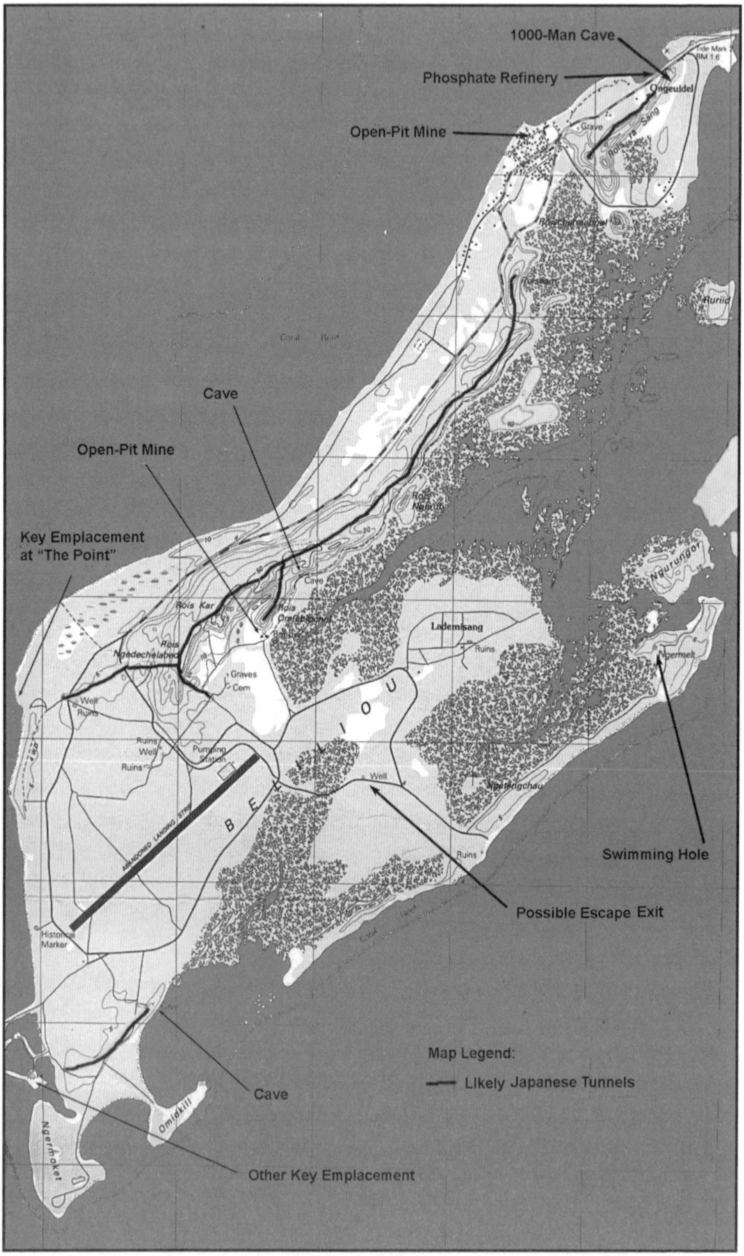

Map 7.1: Likely Whereabouts of Main Japanese Tunnels

(Source: "Map of Peleliu,"Perry Castaneda Collection,from https://upload.wikimedia.org/wikipedia/commons/2/2d/Peleliu_map.jpg,U.S,Geological Survey,public domain.)

More Than One Island Contour Had Helped the Defender

Peleliu's more normal ridgelines had something in common with the little ones. Their similar alignment made them mutually supportable from a greater distance. Following the peninsula's northeasterly flow, all faced one another across a succession of gullies, draws, box canyons, and various declivities. But where those declivities had been closely arrayed and steep-sided, their embedded caves became much like the cliff-side galleries of Biak's Parai Defile.

> The Japanese used this nightmare of crags, pinnacles, and coral rubble honeycombed with natural and man-made caves to bolster their defenses. They were masters at hewing the coral ridges and cleverly camouflaging their positions, from one-man spider holes to enormous four- or five-story caves. Elsewhere on the island, the defenders manned similar, though generally smaller-scale, defensive positions.[21]
>
> — *Naval History Magazine,* October 2019

The Makings of a Perfect MLR?

The nearly parallel razor-sharp ridges in close proximity to one another on Peleliu had created a defense line that was impossible to cross in a hurry. Events at their crest had little effect on cave occupants below. Any American reaching the top of one became susceptible to fire from the face of the next, even when its summit had been occupied by friendlies. (Look again at Figure 7.2.) While mountain-slope-occupying GIs had great trouble spotting any opposition beneath them, they occasionally lowered explosives into a cave mouth.[22]

Because each lengthwise defile between ridgelines undulated, enemy machineguns could be positioned to cover every bottom segment by fire, just as they had the zigzag tank ditch in Figure 5.4. Each section of narrow gorge then became the kill zone for a Japanese ambush.

> [There were] able defenders in underground caves and fortifications within an incredible jumble of ridges and cliffs. Every advance opened the advancing Marines to new fire

> from heretofore hidden positions on flanks, in rear, in caves above or below newly won ground.[23]
> — *Bloody Beaches,* by Gayle
> Marine Corps Historical Center, 1996

Modern-day infantrymen should pay particular attention to this unconventional way of defending broken terrain. Enough to slow down Chesty Puller's 2GW experts, it may have been a major advance in the tactical evolution of warfare. The difficult terrain had been chosen for two reasons: (1) it didn't appear capable of mounting a viable defense; and (2) U.S. supporting arms had difficulty reaching its inner recesses.

Part of the Mountain Had Been Defended Differently

On Japanese-drawn Maps 6.2 and 6.3, the little circles connote strongpoint matrices—defense zones comprised of mutually supporting bastions. Such matrices appear all along the island's northwestern coast, with a few stretching from shoreline to hillmass. But, at the southwestern base of the Umurbrogol appears a "curved line." This line depicts a linear array of emplacements—an atypical way of defending where separate strongpoints had been the rule. So, the parallel crevasses may have been at mountain's bottom, and not high up on its slopes. The caption on an official photograph suggests they were encountered almost immediately upon tackling the mountain. However, this photo appears in a paragraph entitled "Encirclement of the Umurbrogol Pocket."

> Once the troops entered the Umurbrogol Mountain, they found sinkholes and difficult terrain much as pictured here. Japanese soldiers in the caves and heights above could fire at will at the Marines, who were like so many "fish in a barrel." Department of Defense Photo (USMC) 108432.[24]
> — *Bloody Beaches,* by Gayle
> Marine Corps Historical Center, 1996

That same Japanese map shows a strongpoint matrix at the southern end of the Umurbrogol's summit and smaller ones along either side of its bowl region. Together, they could have constituted the final MLR.

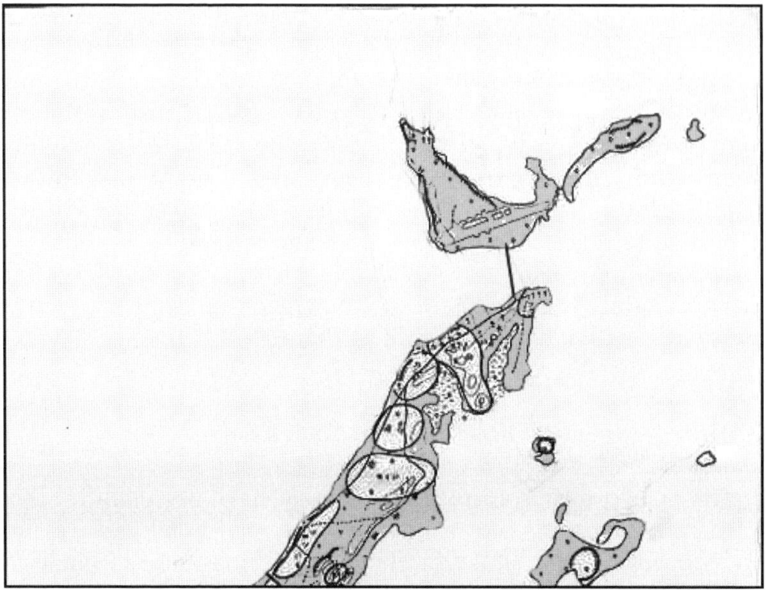

Map 7.2: Japanese Map of Northern Strongpoints

(Source: Japanese Map of Peleliu, History of War website, possibly from captured 31st Army files, at this url: http://www.historyofwar.org/Pictures/pictures_peleliu12.html.)

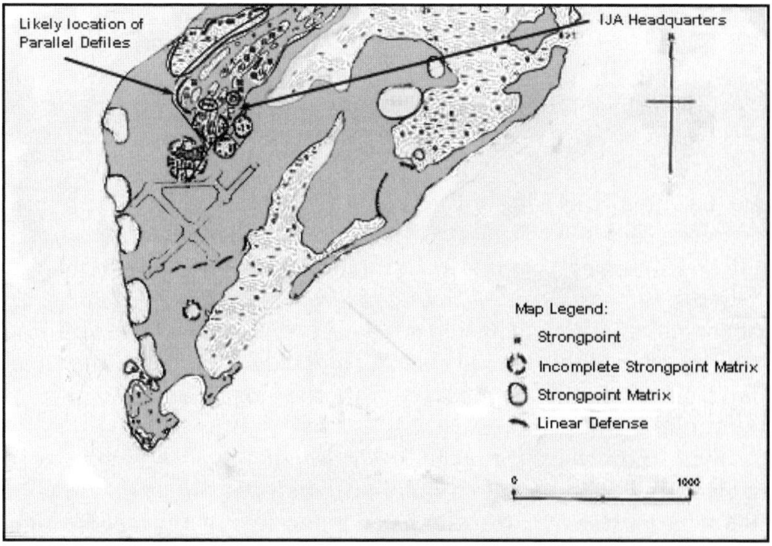

Map 7.3: Japanese Map of Southern Strongpoints

(Source: Japanese Map of Peleliu, History of War website, possibly from captured 31st Army files, at this url: http://www.historyofwar.org/Pictures/pictures_peleliu12.html.)

On the Umurbrogol's forward face had been any number of caves containing an artillery piece. Their relative isolation suggests linkage to a central tunnel system. But, the defenders of the thin back-to-back cracks would have to travel a short distance above ground to reach any sub-surface passageway. Those narrow bands of real estate had not been as well connected as Iwo Jima's thicker ridgelines later in the war. The cliff face Peleliu caves were nearly impossible to reach from the front. Their occupants guarded the forward defile from high overhead and far to one side. They had almost certainly entered their lofty perches from the back via the next uphill declivity. (Look again at Figure 7.2.)

Another Aspect of Those Lateral Crevasses

The Umurbrogol's surface bands must have been similar (but of different origin) to the razor-thin ridgelines in Okinawa's current Northern Training Area. Those on Okinawa were made by roughly parallel streams wearing down the rocky surface to where deep gorges occurred 20-yards apart. But, those on Peleliu had been caused by subterranean pressures cracking the protruding coral. While the fissures on Okinawa led to a lower portion of the watershed, those on Peleliu ran sideways. The rainwater to hit them drained below ground without creating any downhill draws. After dissolving inner-mountain limestone, it may have ended up in a sub-surface flatland stream.

Two Tunnels Had Likely Been Dug Beneath the Coastal Plain

The thin layer of soil over the island's coral shelf made "digging in" all but impossible for U.S. assault troops, whereas Japanese defenders had drilled and blasted out their positions.[25] At two locations along the edge of the airfield clearing had been considerable excavation. Its evidence comes, in part, from the surprise counterattacks they would spawn. (See Map 7.4.)

In all, there were three slightly raised portions of Peleliu's airfield clearing that aerial surveillance had failed to notice amongst the low vegetation. Naturally formed and five to ten meters high, one rise ran southward a short distance from the Point. Though

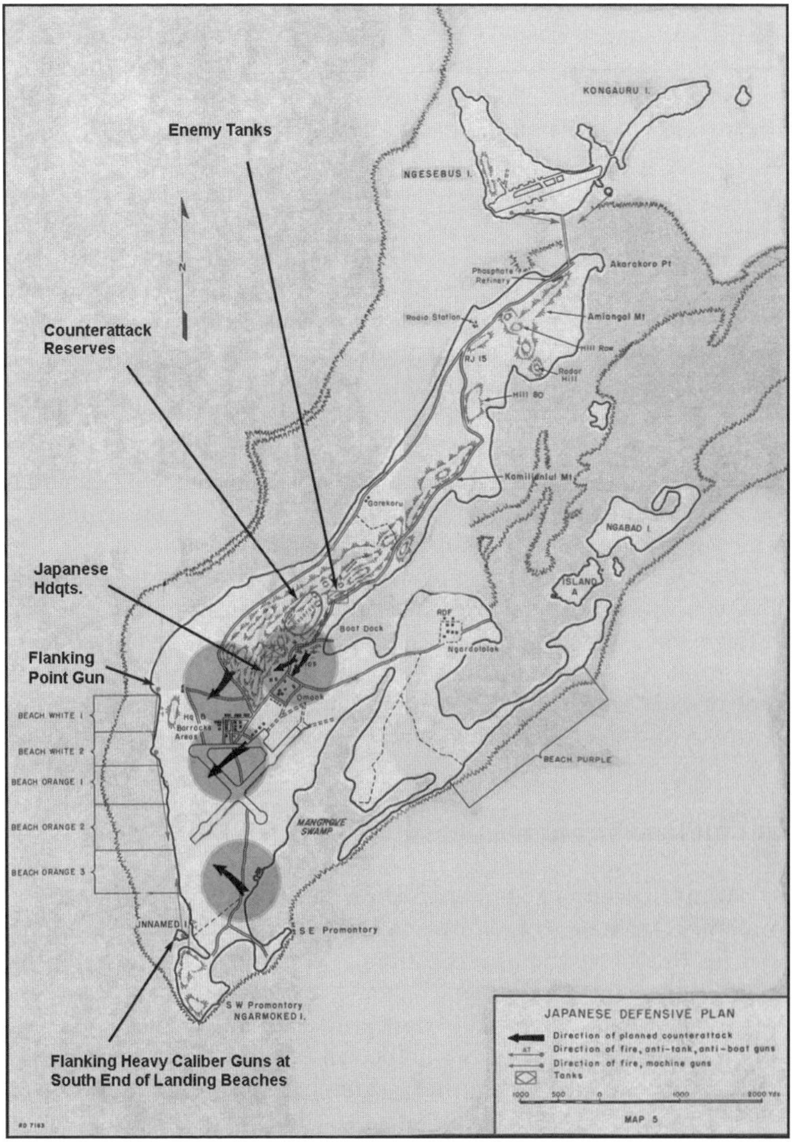

Map 7.4: All Counterattacks Came from Below Ground
(Source: "The Assault on Peleliu," by Maj. Frank R. Hough USMCR [Washington, D.C. Hist. Div., HQMC, 1950], map 5.)

pockmarked with enemy positions, it probably contained no tunnel. Several sources confirm its existence came as a complete surprise to the assaulting Marines.[26]

Much wider, and only slightly higher was the second rise. It stretched from the base of the Umurbrogol to within 300 yards of the Point. (Look back to Map 4.1) Along its southern edge had been many machineguns. And from the end of this squat Umurbrogol finger had issued forth a big Japanese counterattack on the night of D+1, according to Map 5.1. Near that finger's tip was not only a water well, but also a large blockhouse surrounded by tunnel-connected bunkers.[27]

The third raised piece of terrain connected the southern defense-gun promontory with the edge of Peleliu's lagoon. From the top end of that natural mound had originated another big counterattack. Near the source of that attack had been both a strongpoint under construction and a shoreline cave. (Look more closely at Map 7.1 and then at Map 7.4.) That strongpoint may have been the same one K/3/7 ran into on D-Day while trying to reach the O-1 phase line. At its center was a "circular blockhouse."[28] Like its rectangular sister at the tip of the Umurbrogol finger, this blockhouse probably guarded the entrance to a sub-surface passageway.

So, the second and third of these airfield-adjacent features had likely contained a subterranean conduit for the reinforcement or withdrawal of everyone operating those strategically vital promontory guns. Of course, dirt roads also ran most of the way from the base of the Umurbrogol to both promontories. They had provided a way to disguise excavation debris.

Not All Beachfront Bastions Had a Buried Escape Route

Many shoreline defenders had no defiladed withdrawal option. And those at the southwestern end of the island could only go a short way up the above-mentioned tunnel. Then, they had to complete their pullback above ground.

The Japanese who would counterattack the 1st Division formation from the back had either hidden in a sea level cave during the landing or come up this southern tunnel. (Look again at Map 7.4.) Their subsequent attempt to reach the Umurbrogol would occur on the Marines' first night ashore. That's when the right end of the 5th Marines' line came under attack from both north and south.[29]

Full Extent of the Japanese Tunnel System

Besides the lowland passageways already discussed, most of Peleliu's sub-surface conduits had been within the hill-mass itself. Amiangal Mountain sported is own subterranean network. But, the low swampy ground between it and Kamilianlul Mountain had discouraged any below-ground connection. So, most heavily fortified had been the southern Umurbrogol. There, many caves existed at different elevations along either flank—some linked by rainwater runoff to an underground river. A few near the bottom of the mountain may have offered access to foe's original CP and ammunition stores. After all, the 1st Marines did find 35 caves on the lower slopes of the Umurbrogol's southwestern corner.[30] Luckily for the Japanese, the Americans never discovered any tunnel openings inside the natural caves.

That the mountain's interior had been laced with chambers and passageways is a geological certainty whether or not those spaces were later improved. Look back at Figure 3.1 for what existed at the bottom of a Biak sump. Just as on Biak, the Peleliu defenders had taken advantage of this opportunity to travel below ground.

> Peleliu's defenders connected the island's natural caves with an elaborate network of tunnels, which allowed Japanese troops to attack the advancing Marines from anywhere on the island, including from areas just cleared.[31]
>
> — Friends of WWII Memorial

As U.S. forces moved forward, some of Peleliu's flatland defenders pulled back to the Umurbrogol where their Reserves and headquarters had been waiting.[32] Both blockhouses played a role in this fallback. From inside the Umurbrogol, those withdrawing troops could have helped with counterattacks or to reinforce active positions.

> [T]he Island had a bunker system with over 500 caves and tunnels allowing Japanese troops to relocate at will using old mine shafts to store ammo.[33]
>
> — Peleliu Dive Adventures website

The "retreat combat" strategy from Europe was starting to appear in the Western Pacific.

> The Japanese had adopted a new strategy and concentrated [through continual redistribution] their resistance in an extensive network of more than 600 caves and tunnels within the Omleblochel mountains that extended the length of the island.[34]
>
> — National Park Service Study II

Other Sub-Surface Possibilities

There had been a sub-oceanic communication cable between Japanese headquarters on Babelthuap Island and Peleliu.[35] It eventually led to the Japanese CP atop Momiji Plateau—possibly through rainwater runoff channels. So, sub-surface telephone lines may have existed on Peleliu's coastal plain to help coordinate its defense.

Some 10,500 defenders would be killed in the fighting and 200 taken prisoner. Most of those POWs had been forced laborers from Korea and elsewhere in Asia. No one knows how many other human beings were sealed in the caves and never accounted for.[36] Nor do they know how extensive Peleliu's sub-surface network really was.

There's a small sea level peninsula on the eastern side of the island. Near its end is a tourist "swimming hole" that from overhead looks like a tiny sump. This circular hole in the coral shelf contains a mixture of fresh and salt water. The source of that fresh water could be rain that once fell on the central hill-mass. If so, some the Umurbrogol's erosion-created caverns may have had natural flatland exits. And a well along the way could have provided interim access. (Look more closely at Map 7.1.)

> [A] natural hole (with a ladder) where fresh and salt water mixes because it connects underground to the ocean.[37]
>
> — Dolphin Bay Resort brochure

> [U.S.] engineers discovered that productive wells could be drilled almost anywhere on the comparatively low ground [evidence of drainage basins].[38]
>
> — *Bloody Beaches,* by Gayle
> Marine Corps Historical Center, 1996

Even shallow aquifers have feeder channels from somewhere else. Japanese engineers may have expanded a few of those channels to permit secretive messaging and movement during the invasion. Cable and a switchboard were later discovered in a "Naval Communications Cave." And some of the pillboxes had cable ducts in their cement walls.[39]

Additional Lowland Tunneling

The pre-invasion maps of Peleliu reveal several wells at the southern end of the island, and the University of Hawaii confirms a "water catchment" at the edge of the airport. (Look back at Map 4.2.) That catchment was probably the reservoir and pumping station more apparent on colored versions of that map.[40] Pumping stations are for drawing water up out of the ground, so a natural aquifer is more likely than a man-made cistern. No one knows how many rainwater runoff channels may have fed that aquifer. But, below the airfield service area had been 10-20 feet of dry coral shelf. That's evident from the bomb shelter beneath its administration building. So, even where no tiny rises existed on the coastal plain, there had been plenty of dry ground in which to tunnel.

> At the site of the airfield, three major Japanese reinforced-concrete structures remain. . . . The most impressive of these is the two-story administrative building. . . . Steel shutters cover some of the windows. Nearby is a large underground air raid or storage shelter.[41]
>
> — National Park Service Study I

The All-Important Pocket Caves

Last to be captured on Peleliu Island would be the southern crest of the Umurbrogol (a plateau-like region). While the Hawaii University Map 7.5 depicts few caves in this area, they did exist before the fighting.

> The limestone ridges and ravines in the last center of Japanese resistance, "the Umurbrogol Pocket," undoubtedly contain much evidence of the battle for Peleliu. U.S. Marines

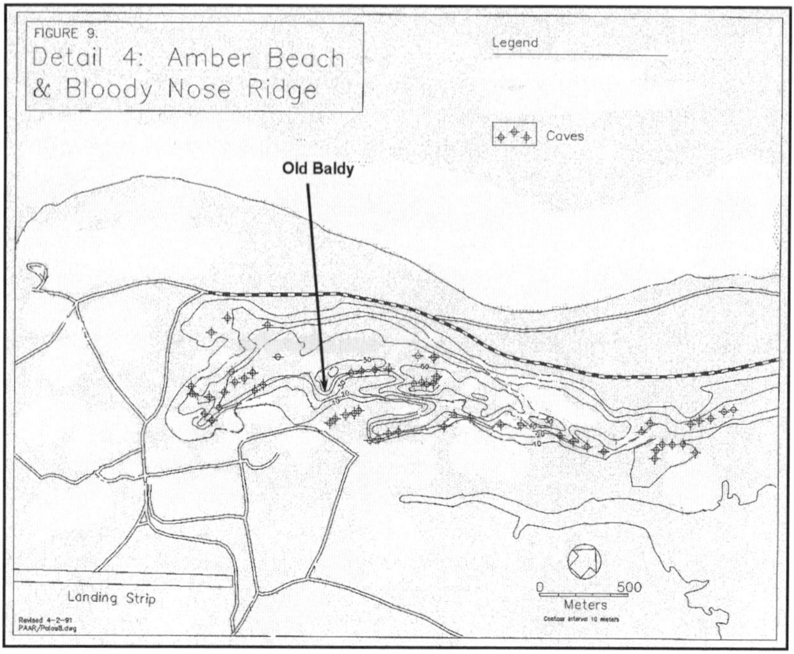

Map 7.5: Still Existing Caves on Umurbrogol Mountain
(Source: "Preliminary Draft -- 1/91 (of Peleliu study)," Botany Dept., University of Hawaii, January 1991, fig. 9.)

> counted over 100 caves that were occupied by Japanese troops in this area. . . . The cave in which the Japanese army commanders committed suicide in a ridge named China Wall has been identified. It is described as being a vertical fault containing three separate caves (D. Colt Denfeld, "Survey of Peleliu," a draft, 1983).[42]
>
> — National Park Service Study I

Degree of Tunneling on the Momiji Plateau

Along both sides of the elongated bowl region at the southern end of the Umurbrogol, Japanese Map 7.3 reveals tiny strongpoints. Sitting over the Reserves cavern, those MLR bastions undoubtedly enjoyed tunnel access to reinforcements. Hill 300 (Old Baldy) contained a vertical shaft and ladder that could only be reached from inside the mountain.[43] This shaft led to an observation post directly above where Maps 4.3 and 4.4 show the original Japanese CP to have been.

Such sophisticated details suggest the entire southern crest of the hill-mass was laced with sub-surface passageways. A U.S. geological survey of the area confirms that some faults in the rock had been sufficiently enlarged through erosion alone. Other faults had human help.

> The same [sub-surface volcanic] pressure [on Peleliu] . . . created innumerable underground faults in the coral limestone; . . . many of which had been further enlarged by [rainwater] erosion into natural caves [and channels], . . . these natural caves [and channels] occurred in all shapes and sizes, and at every conceivable level from the floors of valleys, up the faces of sheer cliffs to the tips of craggy pinnacles. It was the Japanese exploitation of these natural features which caused the fighting in the Umurbrogol Pocket to develop as it did.
>
> The terrain within the Pocket . . . differed only in degree from that in the Kamilianlul and Amiangal systems to the northward: the ridges were higher, longer, and packed more closely together in greater depth. The highest elevation achieved an altitude of only about 300 feet, but that was almost straight up. No caves here were as large and elaborate as some of the engineering feats encountered [at the Pocket].[44]
>
> — *Bloody Beaches,* by Gayle
> Marine Corps Historical Center, 1996

Then, the final struggle for the Umurbrogol Pocket revealed just how heavily prepared the summit of that mountain had really been.

> That feature [the Pocket region] was no level plateau, but a veritable moonscape of coral knobs, karst, and sinkholes. It had no defined ridges or pattern. The sinkholes varied from room-size to house-size, 10 to 30 feet in depth.[45]
>
> — *Bloody Beaches,* by Gayle
> Marine Corps Historical Center, 1996

Sumps are nothing more than big sinkholes. Look back at Figure 3.1 in the Biak chapter to see how hollow the Momiji Plateau may have been. Many passageways crisscrossed the Pocket's various sub-floors. History would record how much Japanese maneuvering they finally permitted.

> The Japanese, even before their counterattack [around 18 October] against Five Brothers, had sent infiltrators into action on the southwest side of the pocket, chiefly by means of tunnels and connecting caves, to points some 300 yards outside the recently established containing lines of the 1st Battalion, 323rd Infantry (1/323 Rpt, 20 Oct-3 Nov 44).[46]
>
> — "Approach to the Philippines," by Smith
> U.S. Army Center of Military History, 1996

8 Rest of the Initial Ambush at Peleliu

- What form did the Peleliu ambushes take on D+2?
- How much maneuvering did the foe do above ground?

Many casualties would occur at the airport clearing.

(Source: "Bloody Beaches: The Marines at Peleliu," by B.Gen. G.D. Gayle, Marines in WWII Commemorative Series, p. 19.)

Something Out of the Ordinary Was Still Happening

Throughout the battle, the well-seasoned Marines would encounter the unexpected. It was as if the Asian foe had studied their background, and then done an assortment of things well beyond their normal frame of reference. Those things normally entailed terrain-associated surprise—the distinguishing factor between 3GW and the more elementary 2GW.

By mixing lethal ordnance with a cultural predisposition toward deception, the opposition had generated a debilitating "fog of war."

Remainder of the First Day in Hell

Until the south end of the Umurbrogol could be occupied, Peleliu's new visitors would be subjected to highly accurate indirect fire. Well-adjusted shells rained down on them from artillery pieces so well embedded in the mountain that no U.S. bomb or naval gun could touch them. And thousands of mere mortals were about to cross the remainder of a huge field that could also be reached by machineguns far to their front, right, and left. (Look back at Map 5.2.) This did not bode well for the Gyrenes, because there was now *Nambu* "chatter" from their rear.

Only Two Ways for That to Happen

The soldiers of the Rising Sun could have moved in behind the American units via tunnel or just waited below ground for them to pass overhead. UW (a subset of 3GW) resists through ambush, and attacks via infiltration.[1] But, the difference between defense and offense often blurs. For example, rear-area snares can be set by opposition sappers. A Guadalcanal trick was about to be tried on Peleliu. Unwelcome strangers appeared behind U.S. lines.

> [T]he tactic of "passive infiltration" (i.e., hiding out in over-run positions in order to be in the rear of U.S. troops) . . . played . . . a prominent part in the operation [in the Palaus].[2]
>
> — *The Assault on Peleliu,* by Hough
> History Division, HQMC, 1950

Some of the trouble came from ground cavities to one side of an advancing American unit.

> 18 September 1944. During the night of 18 September, Corporal Kincaid [E/2/1] volunteered to occupy a sharp crest which afforded the only protection to his company's flank position on a ridge. Late that night, an enemy mortar crew emerged from a cave and attempted to set up their weapon. Corporal Kincaid succeeded in destroying the weapon.[3]
>
> — Navy Cross citation for Cpl. John O. Kincaid

Other problems arose from pillboxes that had been, for some reason, bypassed.

> 15 . . . September 1944. Discovering an active hostile pillbox to the rear of our lines at a time when his squad was separated from the remainder of its platoon during the initial landing, Corporal Nicholas [B/1/5] effectively directed his unit against the enemy position until it was destroyed, thereby eliminating a threat to the successful landings of succeeding waves of our troops.[4]
>
> — Navy Cross citation for Cpl. Clarence Nicholas

But, there was one incident in which enemy soldiers directly in the path of a U.S. unit had initially avoided being noticed. Two young Marines knew what to do about it.

> 15 September 1944. After Japanese troops had infiltrated our lines and set up two automatic weapons which were inflicting serious casualties and hampering the advance of our forces, Corporal Radomski (H&S Co., 1st Engineer Bn.) left the comparative safety of his position and, with another Marine, boldly advanced on the hostile strongpoint in daring defiance of the unceasing barrages of merciless rifle and machine-gun fire to charge the Japanese emplacement.[5]
>
> —Navy Cross citation for Cpl. S.B. Radomski

> [15 or] 16 September 1944. Aware that infiltrating Japanese were inflicting severe casualties and seriously threatening our positions, Private First Class Welch (H&S Co., 1st Engineer Bn), with a fellow Marine, left his protected position and, in the face of heavy machine-gun and rifle fire, heroically charged a hostile emplacement where the enemy had set up two machine guns.[6]
>
> —Navy Cross citation for PFC Charles E. Welch

Later on, 81st Division's afteraction reports would more than once mention Japanese soldiers taking refuge beneath the feet of advancing GIs—later to assail the U.S. support network.[7] And more easily reaching the opposition's command and logistics apparatus is, after all, the purpose of 3GW.

Another Way the Advancing Americans Got Surprised

After being hurt at the beach, U.S. assault troops had incurred a rearward enemy presence. But, that was not the extent of their first-day worries. Within the scrub brush around the airport clearing, well-concealed Marines were being accurately targeted by unseen *Nambu* machinegunners. Those lone assassins had likely fired from distant "spider holes" (pits with camouflaged lids). Unless some Leatherneck noticed the head and shoulders of a leaf-covered shooter, no amount of searching would later divulge that shooter's location. (See Figure 8.1.)

Map 5.1 reveals "spider traps" in 1st Marines' sector right after the landing. 7th Marines would then run into accurate sniping over by the lagoon—in brush thick enough to prevent it.

> As Company K [of 3/7] continued to push through the scrub forest . . . the commanding officer had trouble maintaining contact between his platoons. Not only did the thick undergrowth limit visibility to a few feet, but enemy snipers kept up a heavy harassing fire.[8]
>
> — "Western Pacific Operations,"
> by Garand and Strobridge,
> Historical Branch, G-3 Division, HQMC, 1971

From a line of spider holes, enemy light machinegunners had—in combination—could have severely impeded the 7th Marines' advance. With the help of aiming stakes, their mutual fire would create a kill zone in the bushes. After all, *Nambu* teams had been in outpost foxholes at Bairoko,[9] and all light infantrymen are good at natural camouflage.

Even with single-shot rifles, such a group could have done damage. During the long struggle for Peleliu, lone Japanese riflemen were frequently to defend things from afar. It conserved ammunition.

Specifics of the Enemy Tank Attack

Peleliu's liberators were still under heavy bombardment, when their division formation finally started to cross the huge airfield clearing on the afternoon of the first day. Then, out of nowhere, a well-coordinated tank-infantry came at them from the vicinity of the gravel pit at the tip of Walt Ridge. That's near the mouth of Horseshoe Valley. Yet, Maps 7.1, 6.2, 4.3, and 4.4 all show the enemy tank park to have been some 800 yards farther north at the base of the hill-mass.

Because the whole place had been under aerial surveillance, that armor may have emerged from an enclosed space or tunnel. Big hangars are known to have existed on Peleliu near a hill with caves in it.[10] Those hangers may have been of the terrain-simulating variety—mounds of dirt covered with cement and then excavated. But, for whatever reason, those tanks were first spotted at the edge of the airfield.

> Into this division posture, at about 1650, Colonel Nakagawa [the island commander] launched his planned tank-infantry counterattack. . . . The attack emerged from the area north of the airfield [at the southeastern end of the Umurbrogol] and

Figure 8.1: Nambu Fire from a Spider Hole with Aiming Stakes?
(Source: Courtesy of Cassell PLC, from "World Army Uniforms since 1939," © 1975, 1980, 1981, 1983 by Blandford Press Ltd., Part I, Plate 156; FM 21-76 (1957), p. 70)

> headed south, generally across the front of the 1st Marines' lines on the eastern edge of the airfield clearing. The attack moved directly into the 5th Marines' sector where Boyd's 1/5 was set in, and stretched across the southern area of the airfield. Marines in 2/1 and 1/5 took the attackers under fire, infantry and tanks alike. A bazooka gunner in 2/1's front hit two of the tanks. The commanding officer of 1/5 had his tanks in defilade, just behind his front lines. They opened up on the Japanese armor, which ran through the front lines and virtually into his forward command group. Boyd's lines held fast, taking the attackers, infantry and tanks alike, under [fire] with all available weapons.[11]
>
> – *Bloody Beaches,* by Gayle
> Marine Corps Historical Center, 1996

2/1 was the first Marine unit to engage the westerly headed tanks. Some kept coming, and others turned south. But, the Marines' training would not allow any breach to their lines.

> 15 to 20 September 1944. . . . Lieutenant Colonel Honsowetz's unit [2/1] moved inland. . . . Fiercely counterattacked later in the same day by enemy tanks supported by infantry, his men fought desperately to repulse the attack, destroying nine tanks and annihilating the Japanese troops in their zone of action.[12]
>
> – Navy Cross citation for Lt.Col. R.E. Honsowetz

> During an enemy tank attack on the lines of his company on 15 September, Private First Class Bungard [G/2/1] coolly took a position directly in the face of a tank and destroyed the vehicle with his launcher. When two more Japanese tanks came crashing through the lines several minutes later, he again fired his launcher from an exposed position in front of the tanks and destroyed the onrushing vehicles.[13]
>
> – Navy Cross citation for PFC Robert L. Bungard

> 15 September 1944. When an enemy tank counterattack was launched against his company's lines, Private First Class Fessler [F/2/1 bazooka man] valiantly moved out into the open and placed himself in position to fire his rocket

> launcher at the oncoming tanks. Loading and firing his weapon himself, Private First Class Fessler succeeded in destroying three of the Japanese vehicles.[14]
>
> — Navy Cross citation for PFC Isadore Fessler

The men of 1/5 next made contact with the enemy tank column. Again, some foe armor moved into them. That's when a few of the rank and file did some impromptu tank killing. It demonstrated that authority to take all necessary action had been properly delegated.

> 15 . . . September 1944. When his Battalion was subjected to strong enemy tank attack which threatened the entire line, Lieutenant Colonel Boyd [1/5] directed his troops in repelling the attack and destroying the Japanese.[15]
>
> — Navy Cross citation for Lt.Col. Robert W. Boyd

> 15 . . . September 1944. When hostile forces launched two counterattacks against his zone of action, Captain Holland [B/1/5] courageously led his assault company in destroying several enemy tanks and in repulsing both attacks.[16]
>
> — Navy Cross citation for Capt. John W. Holland

> 15 September 1944. With an enemy tank heading directly for his machine gun position after his supply of ammunition was entirely depleted, Private First Class Coppedge [B/1/5] courageously exposed himself to close-range enemy fire to direct a rifle grenade from his carbine grenade-launcher at the hostile vehicle and, although wounded in both legs by shell fragments, succeeded in disabling the enemy tank and thereby contributed to its subsequent destruction.[17]
>
> — Navy Cross citation for PFC John D. Coppedge

> 15 September 1944. When a hostile tank, which was unable to advance during an intense counterattack against our lines, withdrew in reverse and continued to fire at our defending troops, Private Blakely [demolition squad, Company A, 1st Engineer Bn.] immediately jumped on the enemy vehicle and, knocking the enemy turret gunner to the ground, dropped several hand grenades inside the

> tank, killing the rest of the crew and putting the tank out of operation.[18]
>
> — Navy Cross citation for Pvt. Jay C. Blakely

The Japanese counterattack then reached its final destination—the lines of 2/5. 2/5's attached Shermans engaged their enemy counterparts, and recent enlistees took care of any vehicle that broke through.

> 15 September 1944. Sighting a large formation of enemy tanks rapidly approaching our forces, Platoon Sergeant Miller [Company A, 1st Tanks] immediately maneuvered his tank into the hostile formation and engaged them in action. Then, courageously exposing himself to intense enemy mortar, artillery and small-arms fire to deliver more accurate fire orders to his crew from the turret of his tank, he contributed in large measure to the destruction of seven Japanese tanks before they were able to penetrate our lines and, although wounded by hostile fire, continued in his gallant efforts until all enemy tanks were destroyed.[19]
>
> — Navy Cross Citation for Sgt. James D. Miller

> 15 September 1944. During a hostile tank counterattack, Private First Class Lyth [E/2/5] picked up his light machine gun and two belts of ammunition and, moving directly into the path of an attacking tank, deliberately exposed himself to its fire while he delivered accurate fire into the vision slot of the vehicle, killing its operator and stalling the tank within thirty feet of his gun. When the remainder of the tank crew attempted to escape, he annihilated them with hand grenades, thereby preventing the tank from breaking through his company's lines and inflicting heavy casualties.[20]
>
> — Navy Cross citation for PFC Dale C. Lyth

Though the frontline Marines had done a wonderful job, their exposed position out in the middle of the airfield clearing still seemed a little tenuous. After expecting a suicidal *banzai,* Gen. Rupertus' staff officers may have become overly elated at destroying so many enemy tanks. But, the Japanese knew their armor would be of little use on the mountain and not last long on the plain. Only as bait for a trap did it have much value. They further realized that U.S. forces could not safely cross the massive clearing in broad daylight. So, by expending their armor, they made sure the American formation couldn't reaching the other side on D-Day.

Exploiting the Failed Counterattack

The full expanse of Peleliu's southern plain was still under easy-to-adjust direct fire from heavy cannons on the Umurbrogol's forward face. Japanese mortar observers may have enjoyed the same birds-eye view. Because, about the same time as the enemy tank attack, 3/5 lost its entire command group to a "lucky" mortar barrage.

After so easily defeating the enemy's armor, landing force leaders decided to capitalize on the enemy's failure. They allowed the tail of the U.S. formation to continue northward. Some of the units would have a long way to go in the open, but evening's approach created a false sense of security. (Refer back to Map 7.2 for approximate extent of this first-day advance.)

> 15 . . . September 1944. . . . In the face of concentrated hostile artillery, mortar and machine-gun fire, he [Capt. Holland of B/1/5] gallantly maneuvered his company across fourteen-hundred yards of open airfield to drive the enemy from strong defensive positions at the far side, and successfully held the ground.[21]
>
> — Navy Cross citation for Capt. John W. Holland

> 15 September 1944. . . . [A]fter repulsing a strong Japanese counterattack, Major Gayle [2/5] skillfully seized the critical moment to cross the Peleliu airdrome, personally leading his Battalion in the assault over fourteen hundred yards of open ground in the face of intense hostile mortar, artillery and machine-gun fire.[22]
>
> — Navy Cross citation for Maj. Gordon D. Gayle

> 15 . . . September 1944. Although wounded while leading an assault across the southern edge of an airstrip, First Lieutenant Mallory [E/2/5] steadfastly refused to be evacu-

> ated and, returning to the foremost elements of his company, gallantly led an attack which forced the enemy back another three hundred yards before dark.[23]
>
> —Navy Cross citation for 1stLt. Donald L. Mallory

Somehow, 3/5's companies had gotten separated. But, that battalion was needed to fill the gap between 2/5 and the 7th Marines. So, the 5th Marines CO ordered his Executive Officer, Col. Lewis W. Walt, to take charge of the unit. Upon corralling its companies, Walt extended Division lines to the lagoon and then down to 7th Marines eastward-moving formation. Following all necessary tie-ins around 2230 hours, the 5th Regiment's new position came under counterattack from both north and south.

Then Came D+1

What happened the next morning would defy imagination. Barely visible in Map 4.2 and 4.3 had been two tiny ridgelines: (1) one extending southwest from Umurbrogol Mountain; and (2) the other down near the southern promontory. Between them lay the island's main runways and 2000 meters of coral shelf covered by less than a foot of topsoil. Just north of the airfield were the remnants of its service area. From all three raised areas could potentially come a deluge of bullets. Even .50-caliber machineguns have a range of 1800 meters, and much of the U.S. landing force was still strung out across a veritable "pool table" of unobstructed terrain. That force had achieved the O-1 phase line and was now obligated to reach O-2.

Peleliu's Biggest Blockhouses Were Still Functional

Most guns at the rectangular blockhouse in 1/1's sector and circular counterpart facing 3/7 were still in operation on the morning of 16 September. Only 2000 yards apart, those guns posed quite a threat to the next advance of 2/1 and the 5th Marines. In fact, a map in *Naval History Magazine* shows a nearly linear U.S. formation had pivoted northward with the rectangular blockhouse as its anchor.[24] That means all fire from that gun site would have bored through every composite unit And, though well dispersed, the Marine line had also encountered a wall of lead from the southern blockhouse. (See Map 8.1.)

> Companies K and I [of 3/7] advanced directly east [on 16 Sept.], with L following in reserve. First task of the left flank unit, Company K, was to reduce the fortifications that had held up the unit on the previous day. Aided by point-blank fire from the tanks that paced their advance, the infantrymen quickly seized the barracks area and the three gun positions, but the blockhouse proved to be a more difficult problem. Its five-foot thick reinforced concrete walls withstood direct hits from naval gunfire, 75mm tank cannon, and bazookas; even flamethrowers failed, for one-inch armor plates shielded the blockhouse's gun ports and its two underground entrances. Only after demolition teams worked their way forward under the cover of smoke to lay their charges directly against its massive walls and breached this fortification was forward movement resumed.
>
> The 3rd Battalion gained the eastern shore by 0925.[25]
>
> — "Western Pacific Operations,"
> by Garand and Strobridge,
> Historical Branch, G-3 Division, HQMC, 1971

An Unfortunate Coincidence?

The Marines' lines had been a bit curvy when first established across the airfield on D-Day night, but—in perfect 2GW fashion—they were straightened before departing the next morning. This was to achieve maximum frontal firepower (at the expense of a sideways response). (Look closely at Maps 8.1 and 8.2.) Before long, the formation came under the most concentrated fire those two blockhouses could produce.

> [16 September]. Lieutenant Colonel Honsowetz led his Battalion [2/1] across an open airfield and through a heavily fortified area to neutralize many hostile concrete emplacements and destroy Japanese troops during the advance.[26]
>
> — Navy Cross citation for Lt.Col. R.E. Honsowetz

> 16 September 1944. With only two-thirds of his command left after a day of heavy fighting, Second Lieutenant Hunter [F/2/1] reorganized his men and continued the assault across an airport where, inspired by his daring leadership, the platoon succeeded in destroying three hostile positions which were holding up the advance of the company, and in inflicting heavy casualties on the Japanese.[27]
>
> — Navy Cross citation for 2ndLt. J.B. Hunter, Jr.

More than one way of looking at things played a role in the mishap. The Marines' progress and prospects had looked good to the ship-borne planners on their 1:50,000 maps. But, foe commanders atop the Umurbrogol saw a straight line of fully exposed and slowly moving Americans.

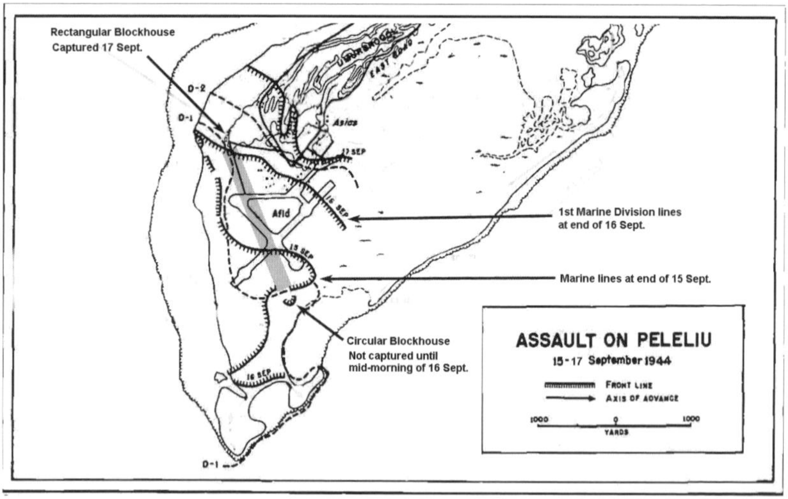

Map 8.1: Straightening Line Crosses FPL Between Blockhouses
(Source: "Bloody Beaches: The Marines at Peleliu," by B.Gen. G.D. Gayle, Marines in WWII Commemorative Series, p. 4.)

> 16 September 1944. The next morning, when his [U.S.] tank-infantry attack was temporarily broken up by direct fire from hostile artillery, he [1stLt. Mallory of E/2/5] boldly joined the forward elements of the company to coordinate their efforts and encourage them to hold their ground until supporting counter-battery fire could be brought to bear on the enemy weapons.[28]
>
> —Navy Cross citation for 1stLt. Donald L. Mallory

> With his platoon pinned down by devastating fire from a hostile pillbox on 16 September, Second Lieutenant Conroy [2/5 platoon leader] unhesitatingly advanced across the open fire-swept area and, personally silencing the enemy emplacement with grenades, enabled his men to continue their advance. When his platoon was again fired upon at dusk the same day, he played a vital part in repulsing the vicious onslaught.[29]
>
> — Navy Cross citation for 2ndLt. Edward Conroy

The Marines' most pressing problem, of course, had been too little cover. Their location was perfectly flat, with only a few shallow shell holes in which to escape harm. When the Marines finally realized a huge ambush was underway, they had little choice but to exit the kill zone.

> We moved at a walk and then a trot, in widely dispersed waves. Four infantry battalions—from left to right 2/1, 1/5/ 2/5, and 3/5 . . . moved across the open, fire-swept airfield.[30]
>
> — *With the Old Breed,* by Sledge (of 3/5)

Even with the increased tempo of movement, the young Americans remained vulnerable to the flying metal.

> We moved rapidly in the open . . . . through ever increasing enemy fire. I saw men to my right and left running bent as low as possible. . . . In many respects it was more terrifying than the landing. . . . We were exposed, running on our own power through a veritable shower of deadly metal. . . .
>
> . . . Japanese bullets snapped and cracked, and tracers

> went by me on both sides at waist height. This deadly small-arms fire seemed almost insignificant among the erupting shells. . . .
>
> . . . How far we had come in the open I never knew, but it must have been several hundred yards. . . .
>
> . . . The attack across Peleliu's airfield was the worst combat experience I had during the entire war. It surpassed . . . all the subsequent horrifying ordeals on Peleliu and Okinawa.[31]
>
> — *With the Old Breed,* by Sledge (of 3/5)

Through no readily comprehensible process, events at the Belleau Wood wheatfield had been partially repeated. Apparently forgotten by those in charge had been how much damage well-positioned machineguns can do at maximum range to an on-line formation. The similarity of this event to the tank ditch ambush of Chapter 5 is no coincidence. Much of the airfield damage had come from the left flank. (See Map 8.2.)

> While the infantry trotted across the airfield, the sound of the wounded and dying was unnerving. "Cries for corpsmen went largely unanswered," wrote Hallas. "Many of the hard-hit Marines bled to death on the hot coral runways before anyone could get to them."[32]
>
> — *Warfare History Network,* 4 January 2019

Some Enfilade Fire Had Been Forestalled

Some of the Umurbrogol finger guns had been close to 2/1's end of the division line. One of that battalion's platoon leaders may have silenced a machine cannon before it could add to the bloodletting.

> 15 to 19 September 1944. Boldly countering the attack of a strong hostile force penetrating our lines on 15 September, First Lieutenant Fournier [G/2/1] repulsed the enemy and resolutely held his lines intact through the night. On the following morning he took a small patrol forward and, locating an automatic gun battery delivering enfilade fire across a strategic airfield, charged furiously, killing twenty Japanese

and knocking out the deeply entrenched fortification which had been holding up the advance of an entire battalion.[33]
— Navy Cross citation for lstLt. Joseph Fournier

September 17th Finally Rolled Around

As the U.S. advance continued on D+2, the left side of the 1st Marine Division formation ran into the most difficulty. While 3/1

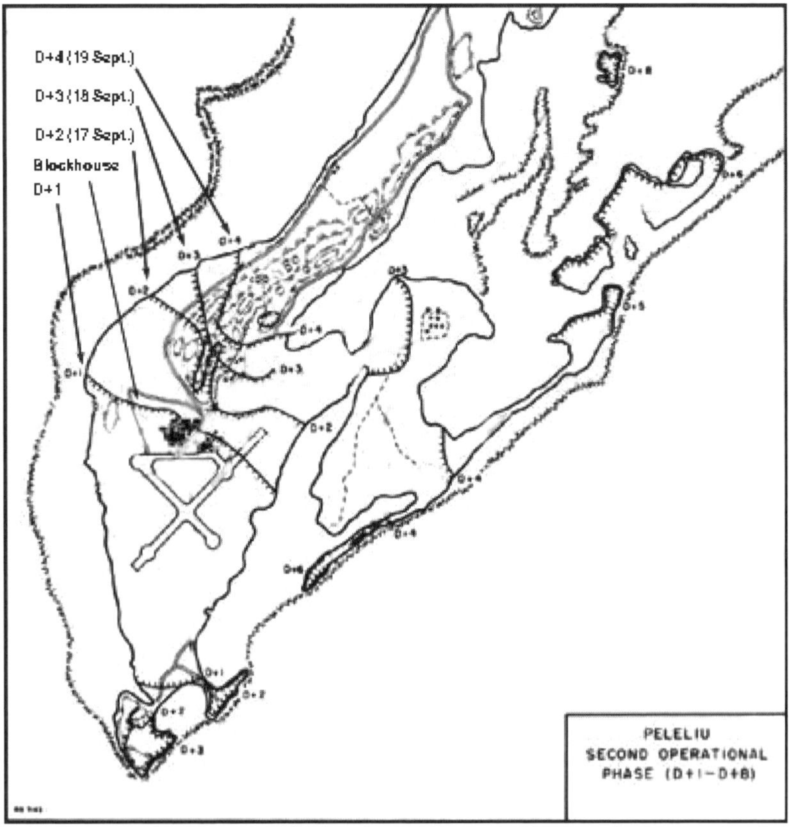

Map 8.2: Location of Division Formation at End of D+1
(Source: "Bloody Beaches: The Marines at Peleliu," by B.Gen. G.D. Gayle, Marines in WWII Commemorative Series, p. 14.)

was still dealing with the seaward end of the Point's fortified swath, 1/1 was finally in a position to do something about the powerful strongpoint cluster at the tip of the Umurbrogol finger. From this location, heavy fire still poured from a large blockhouse with walls four feet thick. Around that blockhouse were a dozen supporting pillboxes, all connected by tunnel. And none of these fortifications had been previously damaged, despite the Navy's claim that all enemy targets in the area had been destroyed prior to the landing. After repeatedly trying to take the blockhouse by infantry assault, the commander of 1/1 reduced it with 14-inch naval shells.[34] Its bunker complex soon fell.

As the 1st Marines Neared the Base of the Umurbrogol

Then, 1/1 moved forward once more. Just before reaching the base of the mountain, it came under intense bombardment from above. But then—with the help of U.S. tank fire—the men of 1/1 (possibly in squad columns on line) began to claw their way up the precipitous incline.

> "Men fought and died along faint paths that ended abruptly in sheer cliffs. They had to turn around and backtrack." By afternoon, on 1/1's left flank C Company had seized Hill 150 and part of Hill 180, while A Company, in the center, had captured Hill 160. [Look back at Map 5.2.] But the gains had come at a huge cost. . . . Thirty-five caves had been captured.[35]
>
> — *Naval History Magazine*, October 2019

The fighting was fierce, and the cost high. At 0600 on D+3, 2/7 would temporarily relieve 1/1 on the mountainside. After reorganizing near the blockhouse for a half a day, 1/1 rejoined the fight for the high ground.[36] (Look back at Map 8.2.)

Rest of 7th Marines Well Occupied Elsewhere

The other two battalions of the 7th Marine Regiment continued to clear the island's southeastern quadrant. 3/7 managed to overcome the heavily fortified area with a circular blockhouse at its center.[37] (Look back at Map 7.3.) With underground entrances, that blockhouse had served more than one purpose. From nearby, a U.S. formation had been attacked from the rear on the first night of the invasion.

9 The Forward Slopes of Bloody Nose Ridge

- Where exactly was Bloody Nose Ridge?
- How many artillery pieces were embedded there?

Chesty's boys had yet to experience true "hell on earth."

(Source: "Bloody Beaches: The Marines at Peleliu," by B.Gen. G.D. Gayle, Marines in WWII Commemorative Series, p. 46.)

Beyond the Flatlands Lay an Impossible Task

While Col. Lewis Puller was later criticized for the "slap-dash" way in which his troops moved up the Umurbrogol, he had only followed established procedure for this stage of the war. Not until the struggles for Iwo Jima and Okinawa would more 3GW-like maneuvers be tried out of sheer desperation.

At Peleliu, 1st Marine Regiment had been given the most challenging initial assignment—the north end of the beach during the landing. It was now under pressure from Division headquarters to quickly climb the island's heavily fortified hill-mass. Against an opponent who had been closely following the tactical developments from Europe, such an endeavor would have its cost in American casualties.

Around the southwestern corner of Umurbrogol Mountain, Japanese tacticians had found a series of lateral cracks in the mountainside. This upper-slope irregularity consisted of a parallel succession of deep, naturally occurring defiles. Their closeness to one another made them nearly unassailable when properly manned. Machinegun cavities within respective cliff faces now insured their mutual support.

Marines attempting to get to the crest of the Umurbrogol's forward slope were thus faced with a closely arrayed group of razor-sharp ridges, each with its own protective crevasse. The landscape alone would have given them pause. But, this nightmarish jumble was also full of nearly inaccessible Japanese positions. U.S. units seldom took a stand on broken terrain, so their leaders didn't understand its defensive potential. Each thin ridge had been turned into a miniature defense belt that could protect the one to its front with unseen fire. (Look back at Figure 7.2.)

The only 2GW way to penetrate such a barrier was for a whole U.S. unit to come up on line with the first crack in the earth and send a tiny team down into it. That team (and any reinforcements) then had to ascend the steep side of the next projection. Trying to secure its crest drew fire from the following cliff face, and so forth. Most of those embedded gun sites had been below the overall slope of the mountain and thus out of sight from the attacking unit. In other words, they were beyond the reach of most small-arms or artillery fire in support of the push.

> [T]he deadly claw-scratch-and-scramble attack of Davis' 1/1 against the Japanese on and in the ridges.[1]
>
> — *Bloody Beaches,* by Gayle
> Marine Corps Historical Center, 1996

> The 1st Marines were trapped in the narrow paths between the ridges, with each ridge fortification supporting the other with deadly crossfire.[2]
>
> — *Wikipedia Encyclopedia,* s.v. "Battle of Peleliu"

Like the cliff-side galleries of Parai Defile on Biak, most of these razor-sharp ridges on Peleliu had loose debris at their base. So, Chesty's options for climbing the Umurbrogol in a hurry had become quite limited, as well as dangerous. When the other Marine regiments finally joined in the fight for the high ground, their casualties would also mount.

> Ravines, which on the map and photographs appeared to be steep-sided, actually had sheer cliffs for sides, some of them 50 to 100 feet high. With nothing else on your mind but to cover the distance between two points, walking was difficult. . . . There were dozens of caves and pillboxes worked into the noses of the ridges and up the ravines. It was very difficult to find blind spots as the caves and pillboxes were mutually supporting. . . . These caves and pillboxes housed riflemen, machine gunners, mortars, rockets, and field-pieces. The Japanese technique was to run the piece out of the cave, fire, [and] then run the piece back in the cave before we could react.[3]
>
> — Gen. O.P. Smith, Asst. Division Commander
> "Personal Narrative"

But First Came the Approach March

After fighting for the Point and Blockhouse, 1/1 had been fully "blooded." Then, the battalion would be so cut up trying to reach the base of the Umurbrogol that it required reorganization. Japanese-drawn Map 7.3 shows a strongpoint cluster at the airfield service area that could have shot at 1/1 from the right and another on the north shore that could have engaged it from the left. All the while, the famous outfit would have endured plunging fire from the mountain. Luckily, 2/1 was now in the process of cleaning out that service area.

> Honsowetz' 2nd Battalion, 1st Marines, moved out in the wake of the preparatory fires across the northwestern portion of the airfield. When the advancing riflemen reached the building area, stiffening resistance from Japanese hiding among the ruins . . . slowed the attack momentarily. Despite heavy casualties, however, the battalion overran

> the enemy defenders in savage hand-to-hand combat and began fanning out toward phase line 0-2.[4]
>
> — "Western Pacific Operations,"
> by Garand and Strobridge,
> Historical Branch, G-3 Division, HQMC, 1971

2/1 would later move around the southeastern corner of the mountain to assail—from the side—the part above the origins of East Road. That finger-like extension had Hill 200 at its end, and the one behind it Hill 210.

How the Southern End of the Mountain Was Defended

Map 4.3 shows a defense line (possibly the multiple cracks) running from just north of the Umurbrogol's southwestern corner to the mouth of Death Valley.

That same map depicts a large strongpoint matrix covering all of Bloody Nose Ridge—everything downhill from the saddle between Five Sisters and Old Baldy. That area included three massive Umurbrogol fingers (from west to east): (1) the ridgeline with Hills 200 and 260 eventually seized by 2/7; (2) 2/1's Hill 210 ridgeline; and (3) the 2/1 ridgeline containing the first Hill 200 and Hill 205. (See Map 9.1.)

The Full-Blown Assault

After the brief respite, 1/1 joined a regimental-sized assault against the Umurbrogol. 1st Marines had just been reinforced by 2/7.[5] So, on D+2 (17 September), the now reinforced 1st Marine Regiment jumped off in the attack—3/1 along the coastal plain, 1/1 on the left side of the massif, and 2/1 at the right of that massif. In reserve at first, 2/7 would eventually fill a gap between 1/1 and 2/1. (Look again at Map 9.1.)

That right-side element would be the first to attack. 2/1's first objective—Hill 200—was on the high ground above East Road.[6] At its base was a swamp.

> On 17 September, he [Lt.Col. Honsowetz of 2/1] braved ter-

> rific fire from enemy mortars, artillery and machine guns to launch an assault against almost impregnable positions dug into the face of steep cliffs dominating a swamp where Company G was pinned down. Determined to relieve the situation, he unhesitatingly made his way across the swamp to the front lines and . . . called for supporting fire and smoke, personally directing the Company in a daring maneuver around the flank.[7]
>
> — Navy Cross citation for Lt.Col. R.E. Honsowetz

Because of enemy artillery embedded in the hillside, 2/1 would require the fire suppression of U.S. tanks and howitzer-carrying amphibians.

> The battalion [2/1] wheeled left and took the ridge under attack, scaling the steep slopes in the face of withering small arms fire and point-blank salvos from mountain guns and dual-purpose pieces which were run in and out of cover. At such close range this fire was extraordinarily accurate. Casualties among the assault troops mounted, and tanks and LVT(A)s moving up in support were knocked out right and left. By nightfall, virtually all of the crest had been taken. The men dug in, as best they could against the fire that now poured on them from elevations farther forward and from a second, slightly higher ridge (Hill 210) which paralleled Hill 200 on the west.[8]
>
> — *The Assault on Peleliu,* by Hough
> History Division, HQMC, 1950

Then, it was 1/1's turn to ascend the southwestern end of the Umurbrogol. That battalion would also suffer, but never lose heart.

> By now, the 1st Battalion, too, was face to face with the ridges. . . . Just across the road began that irregular, unpredictable rising of the land which was such a familiar feature of the landscape of Peleliu's northwestern peninsula. No sooner had 1/1 halted to reorganize than the line was assailed by a fury of fire from the high ground, and casualties began to mount alarmingly.

As quickly as it could be managed, reorganization was completed and the men surged forward again, slugging their way upward much in the manner the 2nd Battalion was storming Hill 200. There was no other way: clawing

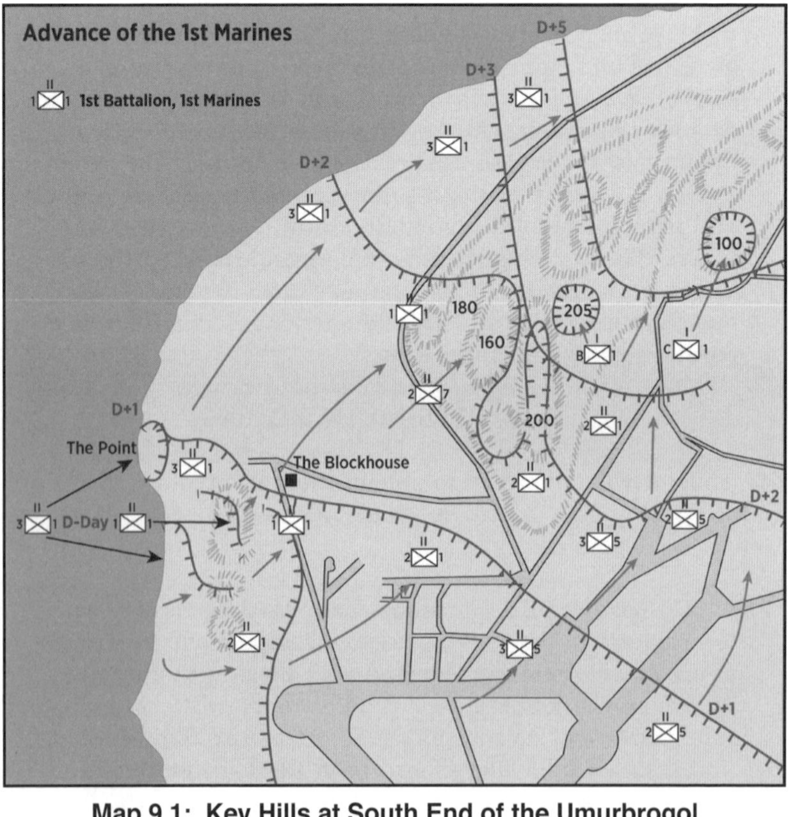

Map 9.1: Key Hills at South End of the Umurbrogol
(Source: Courtesy of "Naval History Magazine," from "Eight Days in Hell," October 2019 issue, © 2019 U.S. Naval Institute/www.usni.org.)

> up and over razor-back crests, shinnying coral pinnacles, plunging down into sheer-sided gullies and ravines, dodging behind boulders. By evening [of 17 September] they had gained the forward slope of the first series of hills and were firmly, if uncomfortably, established." One significant feature of this advance was that, in the course of it, the infantrymen were obliged to knock out 35 separate Japanese-manned caves, existence of which had not been noted in their advance intelligence.
>
> . . . Lieutenant Colonel Davis . . . described it as "the most difficult assignment I have ever seen. . . . Company A depleted itself on the bare ridge on the right as Company C became seriously over-extended on the left and was faltering. Everything was thrown in to exploit Company C's success. Remnants of Companies A and B, [plus] Engineer and Pioneer units were committed to fill the gaps as darkness came. Headquarters personnel were formed into a meager reserve. . . . Company C had moved 800 yards during the day, and we had been able to hold on although the cost was extremely heavy."[9]
>
> – *The Assault on Peleliu,* by Hough
> History Division, HQMC, 1950

Higher Up on the Mountain

Farther up the Umurbrogol, 1/1 may have been the first Marine outfit to come upon the non-assailable cracks in the earth. Hidden along the bottom of each defile were tiny crevices that likely served as portals to the Umurbrogol tunnel system.

> Davis explained: "Men fought and died along faint paths that ended abruptly in sheer cliffs. They had to turn around and backtrack." By afternoon, on 1/1's left flank C Company had seized Hill 150 and part of Hill 180, while A Company, in the center, had captured Hill 160. But the gains had come at a huge cost: 250 casualties. A company commander reported, "We're up here, but we're knee-deep in Purple Hearts." Thirty-five caves had been captured.[10]
>
> – *Naval History Magazine,* October 2019

One battle chronicle has B/1/1 testing the forward slopes of Five Sisters and Old Baldy as early as 17 September, but it probably did so after attachment to 2/1 on the 18th. While Map 4.3 doesn't show defense-line cracks extending that far to the east, B/1/1 would be the first Marine unit to discover steep cliffs at the top of Bloody Nose Ridge.[11]

At some point, a gap had developed between 1/1 and 2/1 that would have to be filled by 2/7. But, its exact location is not clear because of the possible transfer of operational control over units that remained in place.

> The varying fortunes of the different units resulted in an odd-appearing . . . contour to the front line on the evening of 17 September. Because of the inability of the 2nd Battalion (1st Marines) to secure Hill 210, the right of the 1st Marines' zone somewhat resembled a constricted letter "W". And there were two ominous features. While the Marines held a deep salient in enemy territory along the crest of Hill 200, this area after dusk, a gap was created between the flanks of the 2nd and 1st Battalions. This the night-prowling enemy soon discovered and commenced to infiltrate in force. To cope with this situation, it was necessary to commit Company F, 7th Marines, and that unit was obliged to fight its way into position before the gap could be closed. Pressure on Hill 200 was so sustained and heavy throughout the night that at 0200 on the morning of 18 September, the 2nd Battalion was obliged to bring up Company G, 7th Marines, (2/7 "Unit Journal") to bolster the position there.[12]
>
> – *The Assault on Peleliu,* by Hough
> History Division, HQMC, 1950

Sunrise helped to clarify the situation. There had been changes in the 1st Marines' line-up.

> At 0600 the next morning, D+3 [18 September 1944], the 2nd Battalion, 7th Marines, [temporarily] relieved 1/1 on the front line. After falling back to the blockhouse, the 1st Battalion's men . . . [re-outfitted and] reorganized [some more]. That afternoon, B Company received orders to re-

> join the fight. The previous day, 2/1 had seized Hill 200, but now, amid counterattacks, its situation was desperate. To relieve the pressure and establish an observation post, B Company assaulted and seized Hill 205, off the northeastern flank of 200. But its subsequent attack against a complex of rugged peaks was repulsed.[13]
>
> — *Naval History Magazine,* October 2019

After helping 2/7 with an interim objective, A/1/1 joined 2/1 on the original Hill 200. There, it too got the chance to work high up on the mountain, only to suffer the ravages of heat exhaustion.[14] But, Chesty's hard-chargers had made the Japanese commander nervous enough to move his CP to the Momiji Plateau. Deep beneath Old Baldy—just beyond Hill 205—had lain the initial Japanese headquarters.

1st and 7th Marines Now Both Assail the Heights

2/7 and 2/1 again pushed toward the top of the mountain from its southern end. That's when the saddle between Five Sisters and Old Baldy became known as "Bloody Nose Ridge."

> On the 18th, in what the Marine Corps' history of the operation characterizes as "savage and costly fighting," 2 elements of the 1st and 7th Marines managed to advance more than 500 yards in the center, principally along the ridge lines and hills oriented north and south. But the advance during the day was held up at a group of peaks which seemed at first to form a continuous ridge line that was oriented more east and west than the rest of the Peleliu ridges. Soon, the name Five Sisters came to be applied to a cluster of peaks forming the western side of the terrain feature. A towering hill at its eastern extremity, separated from Five Sisters by a saddle, was designated Hill 300, or Old Baldy. With the 7th Marines held up in the center, at Five Sisters, the units on both sides of the central ridges halted so as not to create exposed flanks.[15]
>
> — "Approach to the Philippines," by Smith
> U.S. Army Center of Military History, 1996

But, this U.S. push was through convoluted terrain. Every advance had been by a Marine element small enough to fit into the available compartment. Of course, parent-unit leaders were nearby.

> Although wounded on 18 September, Sergeant Gerycz [of F/2/1] steadfastly refused to be evacuated and continued to lead his squad until ordered to the rear by his Platoon Leader.[16]
>
> — Navy Cross citation for Sgt. Steve Gerycz

> [H]e [1stLt. Fournier of G/2/1] pressed through to the high ground overlooking the airdrome, seizing the position despite heavy resistance and holding it against desperate counterattack. Continuing his relentless drive into the fourth day [19 September], he persistently sought out and destroyed obstinate gun emplacements. When his battalion was again pinned down by a merciless barrage on 19 September, he braved the shattering fusillades to work forward with five others and succeeded in neutralizing the hostile firepower. . . . [T]hen, reorganizing his seriously depleted force, broke through stubborn enemy defenses on a high ridge. . . . Struck down by a burst of machine-gun fire while leading the final assault, [he finally desisted].[17]
>
> — Navy Cross citation for 1stLt. Joseph Fournier

The Attack's Resumption

While other parts of the regiment climbed the Umurbrogol, 3/1 worked to clear the coastal plain over to where the West Road ran northward. There, a big enemy strongpoint likely encompassed a few mountainside caves. (Look back at Maps 7.3 and 7.5.)

Atop of Bloody Nose Ridge, A/1/1 was about to run into the same high cliff that had given B/1/1 fits.

> At 0700 on 19 September, the attack was resumed along the entire 1st Marines' front. Colonel Puller had received no new directives; his regiment still had the assigned mission of seizing the high terrain up to phase line O-2 [and O-3]. The 3rd Battalion, unmolested except by

snipers, moved up the western coastal flats for some 400 yards before halting in order to retain contact with neighboring units advancing more slowly over the ridges. Here, 2/7, still under operational control of the 1st Marines, overcame stiff resistance to seize the forward slopes of Hills 200 and 260 [just west of the 2/1 ridgelines], for a day's gain of 300 yards. Company A of 1/1 passed through to press the attack, but it ran headlong into a sheer 150-foot cliff which, coupled with heavy enemy fire, stopped the assault cold. Only six men out of the entire rifle company managed to regain 2/7's lines without either being killed or wounded.

It was the 2nd Battalion [of the 1st Marines], however, that first tested the strength of Colonel Nakagawa's final bastion. After a 500-yard advance in the face of increasing enemy resistance, the foremost assault units encountered the same foreboding hill mass that had blunted the attack of Company B the previous day. This dominating piece of terrain became known as the Five Sisters, because it contained five peaks; they averaged 250 feet in height and were separated from each other by steep cliffs. The southern face was at first dubbed "Bloody Nose Ridge" by the Marines. No sooner had the battalion consolidated its forward positions than it launched a full-scale assault directly at the forbidding height.

Preparatory air and artillery strikes thoroughly plastered both the forward and reverse slopes of the hill mass, while all tanks and mortars attached to 2/7 were brought over to support the all-out effort of Honsowetz's battalion [2/1]. Rushing forward in small groups to minimize casualties from the terrific enemy fire, the Marines grimly fought their way ahead, and their Shermans, mortars, and LVT flamethrowers ventured as far forward as possible to provide direct fire support. Despite the vigor and determination with which the riflemen pressed the assault, it collapsed completely by noon. Even the most pessimistic Marine present there that day did not dream that the defenders of the Five Sisters would frustrate all attempts to storm them for over two months.[18]

— "Western Pacific Operations,"
by Garand and Strobridge,
Historical Branch, G-3 Division, HQMC, 1971

A different account of the action makes clear what existed atop Bloody Nose Ridge. An MLR had been established along a high cliff. This MLR would permanently form one side of the Umurbrogol Pocket.

> On the right [up the Hill 200 ridgeline], 2/1 slugged its way forward under generally similar conditions [as 2/7] until about noon, when it was brought up short by the same obstacle which had repulsed Company B's abortive effort the previous afternoon: the Five Sisters. Although there was no way of knowing it then, these [U.S.] elements had reached the southern face of what was to prove the final pocket of Japanese resistance on Peleliu.
>
> This curious formation, in contrast to most of the ridge topography, was oriented generally transverse [crosswise] of the island. Thus, in effect, it formed a wall barring passage northward, its southern face almost sheer. Even if the summit could be reached, working along it from peak to peak would be next to impossible owing to the precipitous drops between them.[19]
>
> — *The Assault on Peleliu,* by Hough
> History Division, HQMC, 1950

There had been a sizeable defile at the base of that cliff at the summit of Bloody Nose Ridge.

> Concurrently with the action in the eastern zone, 2/7, tying in on the left of 2/1, slugged its way through the jumbled high ground in an effort to draw abreast of the western flank of the Five Sisters. Some progress was made, but as a flanking movement the attempt was fruitless in that Japanese resistance kept the assailants from reaching any commanding ground from which to bring supporting fires to the assistance of 2/1's attack. Company A, 1st Marines, passed through elements of 2/7 in a turning movement from the west with 56 men in assault. These gained a little ground along a ridgetop which was swept by machine-gun fire, only to find themselves suddenly stopped by a sheer drop of 150 feet. Only six of the Marines who participated in this fight were still unwounded when they finally got

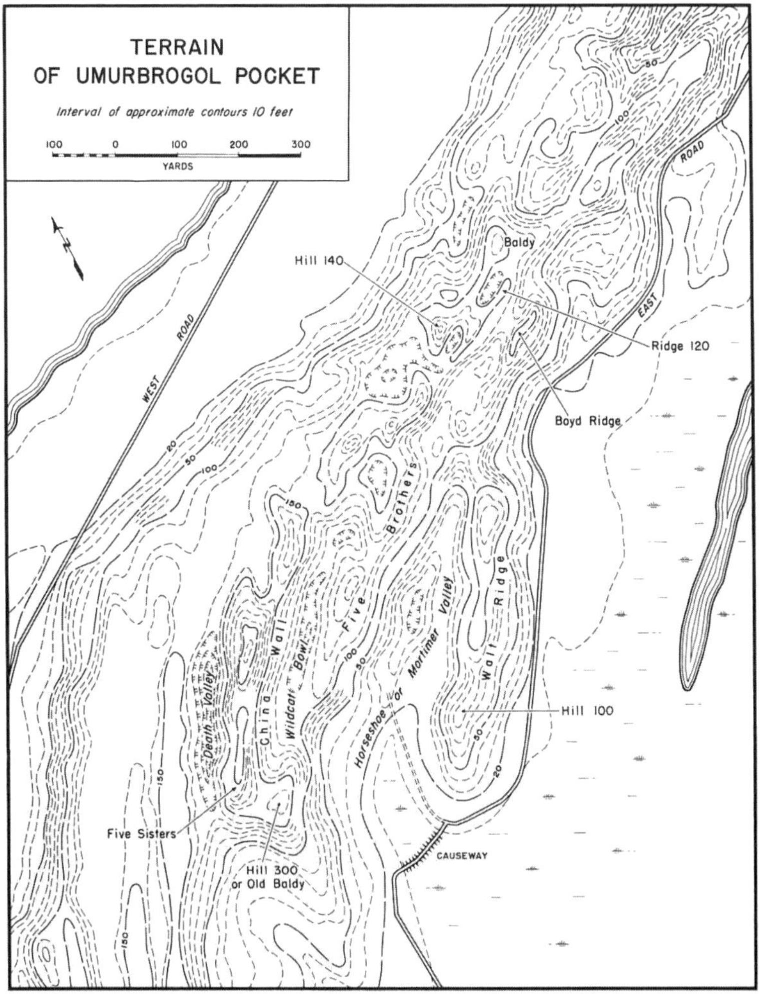

Map 9.2: The Miniature Canyon South of Five Sisters
(Source: "The Approach to the Philippines," by Robert Ross Smith, U.S. Army in WWII Series [Washington, D.C.: Ctr. of Mil. Hist., 1996], map 21.)

back to 2/7 lines.[20]

— *The Assault on Peleliu,* by Hough
History Division, HQMC, 1950

As 2/7 personnel moved eastward, Death Valley first stood in their way. Then, a deep moat-like feature came into view on their left that protected Five Sisters from the Hill 260 finger they were on. All five peaks had other steep cliffs between them.[21] (Look closely at the bottom of Map 9.2.)

> In the center, 2/7 advanced slowly almost due east, battling sheer cliffs as well as heavy fire, and by mid-afternoon Company F, on the right, succeeded in gaining the crest of a ridge (then designated Hill 260) facing the Five Sisters from the [south]west across the mouth of [a] narrow, steep-sided declivity."[22]
>
> — *The Assault on Peleliu,* by Hough
> History Division, HQMC, 1950

Prearranged Schedules Do Have Their Limitations

It had been initially estimated that Peleliu could be taken in three to five days,[23] and the various phase lines named accordingly. But, in trying to reach the O-3 phase line by the end of the third day ashore, the two Marine regiments had taken on too large a challenge. There was nothing wrong with their proficiency, just in how well fortified the most difficult terrain in the Western Pacific had become. Such an area could not be rapidly steamrolled. (See Map 9.3.)

The terrain atop the Southern Umurbrogol had been precipitous and covered with sinkholes. Until its sub-floors could be secured, there was no way to consolidate the surface expanse. So, an envelopment up Walt Ridge seemed the only way for the Marines to achieve the expected progress.

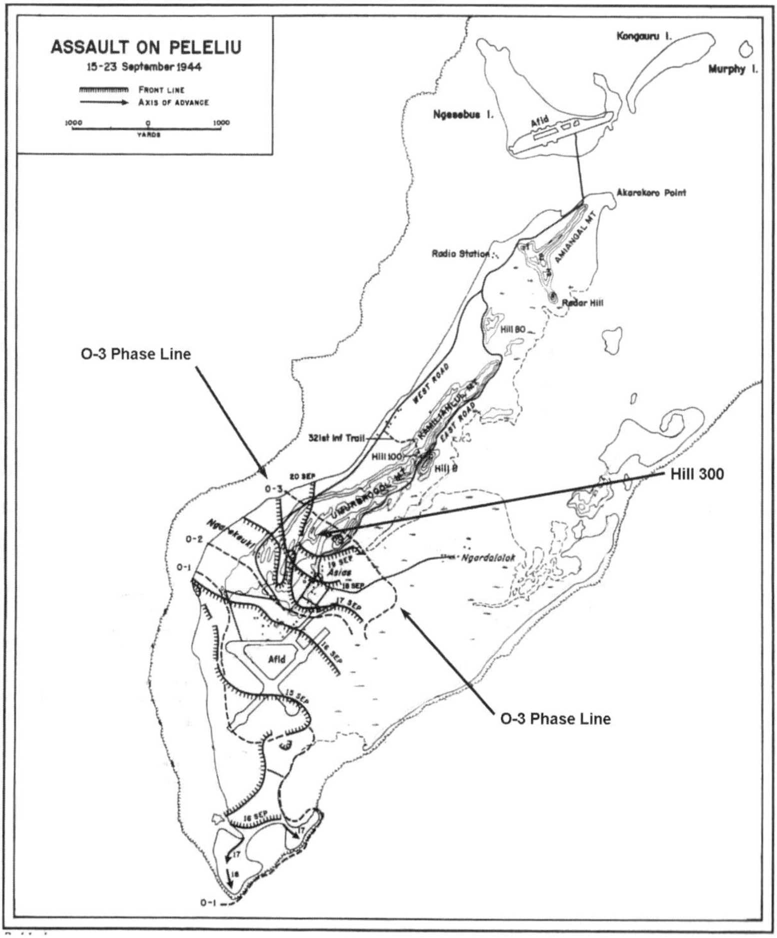

Map 9.3: The Overly Optimistic O-3 Phase Line
(Source: "Blooly Beaches: The Marines at Peleliu," by B.Gen. G.D. Gayle, Marines in WWII Commemorative Series, p. 4.)

10 The Attempt to Climb Walt Ridge

- Why so much difficulty with Walt Ridge?
- Did it contain a tunnel?

More enemy kept appearing at south end of Walt Ridge.

(Source: "Bloody Beaches: The Marines at Peleliu," by B.Gen. G.D. Gayle, Marines in WWII Commemorative Series, p. 17.)

2/1's Next Project

While 2/7 struggled with the central face of the Umurbrogol, 2/1 now tried to ascend the big mountain through a different route—Hill 100 and the rest of Walt Ridge. To help, it had several 1/1 companies attached. (See Map 10.1.)

> Similar action continued the next day [on 19 September], when an effort to take Five Sisters proved abortive. One element of the 1st Marines pushed up East Road through the village of Asias and along the eastern base of a long hill

> known at Walt Ridge. By the morning of the 20th, Company C, after action on Hill 100 at the south end of Walt Ridge, was reduced to 16 men. Companies F and G, hard hit in the center, had to be combined with a squad of men from the 4th War Dog Platoon to form one understrength company. Company A, fighting over high ground on the 1st Marines' left, was reduced by evening on the 19th to six men who had not been wounded or killed.[1]
>
> — "Approach to the Philippines," by Smith
> U.S. Army Center of Military History, 1996

The First Attack on Hill 100 and What Lay Beyond

Hill 100 was next to the valley in which the enemy's tank attack had possibly emerged. Captain Everett Pope and his Company C would succeed in climbing the end of the new Umurbrogol finger, but then be confronted by a slightly higher portion. After running out of ammunition during the night, Pope and his me needed hand-to-hand combat. At dawn, the few men left had to evacuate the hill.

> But, for the few men up on the higher ridge—mostly from C Company, 1st Battalion—it was far worse. All through the night we could hear them screaming for illumination or for corpsmen, as the Japs came at them from caves which were all around them on the hillside. Men were hit up there and we could hear them crying and pleading for help, but nobody could help them. . . . The cries of Americans and Japanese were all mixed together.[2]
>
> — "Battle of Peleliu," *Facts and Details*

Had There Again Been a Difference of Perceptions?

The 1st Marine Division commander was justifiably worried about the 25,000 Japanese soldiers on Babelthuap reinforcing Peleliu. And while working off 1:50,000 maps, his staff officers had lost track of conditions on the Umurbrogol. To them, 100 yards "as the crow flies" seemed the length of a football field. But, for the troops

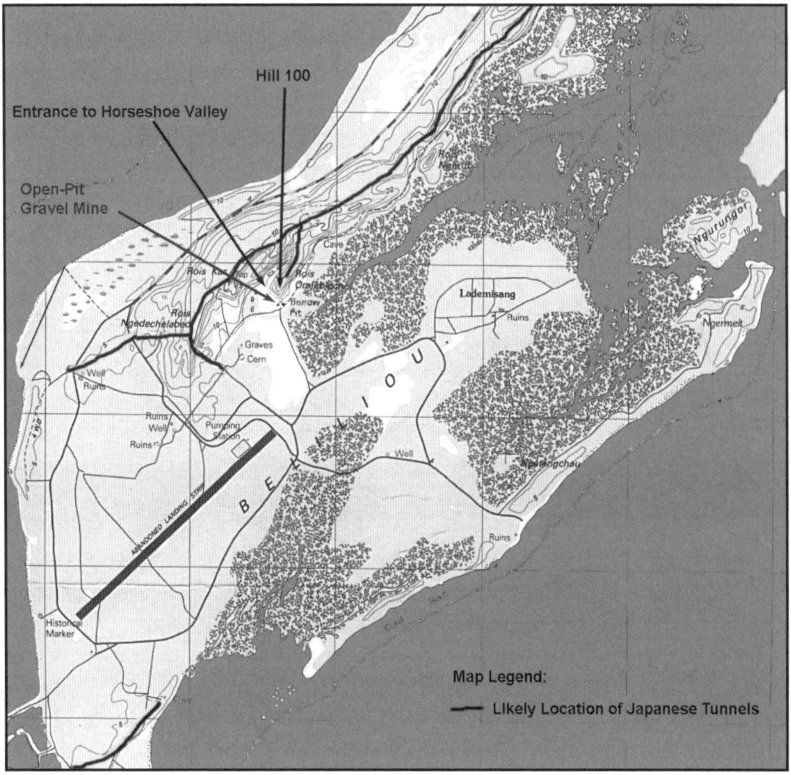

Map 10.1: Hill 100 and Horseshoe Valley

(Source: "Map of Peleliu," Perry Castaneda Collection,from https://upload.wikimedia.org/wikipedia/commons/2/2d/Peleliu_map.jpg,U.S,Geological Survey,public domain.)

on the ground, that same field had been covered with major ups and downs. And the whole southern end of the mountain seemed laced with sub-surface passageways. The enemy could suddenly appear from almost any direction. That's why Col. Puller maintained a reserve component behind every attacking unit and kept Division apprised of the worsening microterrain. But, the headquarters elements of this era were big-picture thinkers. And, upon the advice of these staff officers, Gen. Rupertus urged Chesty Puller to "maintain [the] momentum."[3]

While history has blamed this tactical misconnect on the poor guys in charge, the actual fault lies in the upwards-communication deficit that still exists within most U.S. infantry outfits. Its source is the dramatic difference in scale between "top-down planning" and "bottom-up execution." While crucial gains across difficult terrain may be measured in feet, officially acknowledged progress requires "hundreds of yards."

Details of the Fighting for Hill 100

Early on September 19th, B/1/1 became attached to 2/1. Captain Pope's C/1/1 had already been directed to seize Hill 100. Its parent finger would allow the Marines to outflank the Southern Umurbrogol. Hill 100 also dominated the East Road running northward alongside Peleliu's central hill-mass.

Company C spent the afternoon fighting across swampy ground to the base of Hill 100. Three Sherman tanks lent support, but two slid off the causeway. At 1700, Company C's survivors stormed the height, but discovered higher ground behind it.[4]

> 19 September 1944. Attacking with his squad across a fire-swept causeway behind a platoon of tanks, Sergeant Bartlett [of C/1/1], upon reaching the causeway, noted that the tanks were unable to cross. With his company momentarily halted on the edge of the lagoon and immediately subjected to heavy concentrations of enemy fire, he quickly rose up under the hail of small-arms fire and started across the fifty-yard stretch of exposed causeway. Following his gallant example, his squad and the balance of the company followed him through the fire despite severe casualties. Upon reaching the other side of the lagoon, Sergeant Bartlett quickly reorganized his squad and, after renewing the attack to seize his objective, organized and held the ground against vigorous enemy counterattacks until he was seriously wounded.[5]
>
> — Navy Cross citation for Sgt. John W. Bartlett

Map 7.5 shows the south end of Walt Ridge to have had a string of fortified caves at its base. And, only narrow Horseshoe Valley separated it from the full mass of the Umurbrogol.

> Company C began this action with 90 men. About noon [on the 19th], Captain Everett P. Pope, its commanding officer, was ordered to seize Hill 100, a steep, seemingly isolated elevation which dominated the East Road and the swampy low ground to the battalion's right front. Company C approached through the swamp and reached the road at the base of the hill, where two strong pillboxes were discovered. The men were in the process of assaulting these when they were suddenly pinned down by machine-gun fire at a range of about 50 yards from across a small pool on their right. Unable to get at these assailants, the captain obtained permission to withdraw the way he had come, pass to the left of the main lagoon and attack up the road with the support of tanks.
>
> The road at this point angled sharply to the east, traversing the main lagoon, or sink hole, on a causeway across the mouth of a wide draw which was to gain an evil reputation under the name of Horseshoe Valley, or just "the Horseshoe." Beyond that, it skirted the base of the objective hill past the positions where the company had previously been pinned down, then angled to the northeastward.
>
> The withdrawal was effected successfully, but it was late afternoon before the new advance began. The first tank to attempt the causeway slipped over one side and became immobilized. A second tank, in an effort to extricate it, slipped off the other side, thus blocking the narrow approach to further supporting arms. Nevertheless, Company C crossed it in squad rushes, paused momentarily at the foot of the hill, then assaulted with only mortar and machine-gun support. The men's rush carried to the summit, where they were disconcerted to discover that, instead of having seized the crest of an isolated hill, they had merely gained the nose of a long ridge where their own position was dominated by a higher knoll only some 50 yards to the front (Ltr. Maj E. P. Pope to CMC, 8Mar50).
>
> The Japanese reacted with promptitude and vigor.

On that open ridge crest, the men of Company C were exposed not only to fire from the commanding ground to their front, but crossfire from the parallel ridge that rimmed the Horseshoe on the west. . . . Counterattacks were incessant throughout the night and produced some savage close-in fighting that degenerated into a matter of fists and rocks and throwing the enemy bodily over the cliffs. But at dawn, with only 15 men and one officer left, Captain Pope was ordered to withdraw to the main line, a movement again effected successfully through the swamp."[6]

— *The Assault on Peleliu,* by Hough
History Division, HQMC, 1950

More on That Perimeter Fight

What happened within the lines of Company C that night are now legendary. Company B would be later involved in the fighting.

Pope and his men had set up a protective perimeter that the Marine captain later recalled was about the size of a football field [probably not that big]. Company B [actually C] did have some mortar support, but since they were surrounded they only had what ammunition they originally took with them. There were about two dozen Marines [eventually] left to hold Hill 100.[7]

— *Warfare History Network,* 10 August 2016

The sun went down, beginning a new phase in the battle for the hill. The Japanese launched a series of counterattacks that lasted all night. It was a time when the sounds of men screaming for help or crying in pain mingled with the staccato chatter of gunfire, a terrible cacophony that echoed and re-echoed through the darkness.[8]

— *Warfare History Network,* 10 August 2016

As twilight fell, the Marines took [any] cover they could among the jumbled rocks. The Japanese went for Pope's men after dark, and they kept coming. . . . [T]wo Japanese

> suddenly materialized near the position defended by Lt. Francis Burke . . . and Sergeant James P. McAlarnis.[9]
>
> — *Naval History Magazine,* October 2019

> Fighting became hand to hand. At one point, Lieutenant Francis Burke and Sergeant James McAlarnis were attacked by some Japanese soldiers at close quarters. One Japanese thrust a bayonet into Burke's leg, but the lieutenant pummeled his assailant with his fists. While Burke was occupied, McAlarnis used a rifle butt on a second attacker. The two Marines then pitched the Japanese bodies down the slopes.[10]
>
> — *Warfare History Network,* 10 August 2016

Hill 100 was more heavily fortified than anyone had thought possible. And heavy cannons covered it from across Horseshoe Valley.

> [O]n the night of 19 - 20 September 1944. Assigned the mission of assaulting Hill 100, a heavily-defended coral nose studded with caves and concrete pillboxes, Second Lieutenant Burke [of C/1/1] gallantly led his platoon in a furious attack and reached the summit of the hill. Immediately coming under a heavy rifle caliber field piece which caused numerous casualties, he quickly consolidated the nine remaining men of his platoon into a defense of the forward portion of the hill. Although he received a severe bayonet wound in a hand-to-hand encounter with an enemy soldier, he resolutely remained in his exposed position and continued to lead his men in desperate combat with numerically superior attacking Japanese forces. At dawn, with his automatic weapon out of action, his ammunition exhausted and his position raked by heavy machine-gun fire, he was ordered to withdraw his men to a new position.[11]
>
> —Navy Cross citation for 2ndLt. Francis T. Burke

> [O]n 19 and 20 September 1944. With his platoon assigned the mission of seizing the strongly defended summit of Hill 100, a coral nose studded with caves and concrete emplacements, Platoon Sergeant McAlarnis [of C/1/1] boldly led elements of his unit through intense hostile fire to the top

> of the hill and then assisted his Platoon Leader in deploying his men despite fire from a fieldpiece on an adjoining ridge [Walt Ridge]. On several occasions, he engaged the enemy in fierce hand-to-hand combat and, when his Leader was wounded and pinned to the ground by a bayonet in the hand of a Japanese, boldly rushed to his comrade's aid. Throwing himself upon the enemy, he beat him into submission and hurled him over a cliff. When his platoon's position became untenable the following morning, Platoon Sergeant McAlarnis remained until the last man had withdrawn.[12]
>
> — Navy Cross citation Plt.Sgt. James McAlarnis

The Hill 100 assault force had done well in the face of overwhelming odds. Only later would anyone realize why the few initially present enemy soldiers seemed to multiply.

> Dawn came, probably the most beautiful morning the survivors had ever seen. Toward the end of the fight Company B's [C's] ammunition was so low the men were pitching rocks at the enemy—not that they expected to hit anybody hard enough to do damage. The captain later explained that the Japanese would delay an attack for a moment, thinking the rocks were grenades. To keep the enemy off balance the Marines occasionally threw a real grenade, part of their ever-dwindling supply.[13]
>
> — *Warfare History Network,* 10 August 2016

> At last, even courage had to give way before superior numbers. The Japanese were getting ready for yet another attack, but Pope received orders to withdraw from Hill 100. In the end only nine men, Pope included, successfully made it back to American lines.[14]
>
> — *Warfare History Network*, 10 August 2016

C/1/1's Hillside Stand May Have Had Nearby Support

There's evidence of a Marine reconnaissance unit attempting to move to ground higher than Hill 100, and thus farther up Walt Ridge.

> It had not been Colonel Honsowetz's intention to send Company C on this mission alone, and the Reconnaissance Company had been ordered to support it on the right [farther along Walt Ridge]. However, Captain Pope had already carried the summit before the 2nd Platoon of the latter unit, supposed to tie in here, could get into position to assault, and "There is some doubt as to whether Captain Pope realized this platoon was supporting him during the entire night to follow" (Ltr. 1stLt R. J. Powell, Jr., to CMC (undated) received 21Mar50). With no flank contact with the troops atop the ridge, the platoon also received some rough treatment throughout the night, and at dawn fell back to the cover of the swamp's edge before an attack in force. For some reason, the Reconnaissance Company did not receive the withdrawal order until half an hour after the retirement of Company C and came under additional fire from the reoccupied positions on the ridge crest. Falling back at last via the swamp route, the survivors reached the main line safely, where soon afterward the depleted unit was relieved.[15]
>
> — *The Assault on Peleliu,* by Hough
> History Division, HQMC, 1950

> [Around] 18 September 1944. Leading his platoon in an attack against a group of enemy pillboxes in this area, Gunnery Sergeant Tracey [of Div. Recon. Co.] launched an attack with hand grenades to destroy three machine guns. When a hostile hand grenade landed in the immediate vicinity of his squad leader, endangering his life, he immediately rushed to the man and shielding him with his own body, saved the Marine's life, although he himself was severely wounded by the exploding grenade.[16]
>
> — Navy Cross citation for Gy.Sgt. Warren Tracey

But, how much that recon outfit had actually helped Company C is not clear. Pope's unit later took fire from where the recon unit was headed before meeting so much resistance.

> Subjected to point-blank cannon fire which caused heavy casualties and badly disorganized his company while assaulting a steep coral hill, Capt. Pope [of C/1/1] rallied his men and gallantly led them to the summit in the face of

> machinegun, mortar, and sniper fire. Forced by widespread hostile attack to deploy the remnants of his company thinly in order to hold the ground won, and with his machineguns out of order and insufficient water and ammunition, he remained on the exposed hill with 12 men and one wounded officer determined to hold through the night. Attacked continuously with grenades, machineguns, and rifles from three sides, he and his valiant men fiercely beat back or destroyed the enemy, resorting to hand-to-hand combat as the supply of ammunition dwindled, and still maintaining his lines with his 8 remaining riflemen when daylight brought more deadly fire and he was ordered to withdraw.[17]
>
> — Medal of Honor citation for Capt. Everett Pope

Needless to say, most of that terrible night's exploits were by rank and file Marines. They do, after all, constitute the majority of any infantry outfit.

> [O]n the night of 19 - 20 September 1944. One of five remaining men to gain the forward strong point on the crest of a hill late in the afternoon following a successful assault on the high coral ridges by the remnants of his company, Private First Class Packanik [of 1/1] courageously assisted in defending the ridge against violent Japanese counter attacks throughout the night, repeatedly exposing himself to intense hostile fire in order to direct his own fire more accurately and at one time engaging five of the enemy in hand-to-hand combat when they reached his position. Undaunted in his efforts although his ammunition was exhausted shortly before dawn he tenaciously continued his heroic resistance furiously hurling rocks and stones down upon the enemy and fighting hand-to-hand until mortally wounded during the fierce action.[18]
>
> — Silver Star citation for PFC Walter Packanik

When morning finally came, the enemy mounted a counterattack in greater numbers than previously noticed. That attack must have come from below ground.

> As dawn approached, C Company was down to about a dozen men and was running out of ammunition. "Pope re-

> ceived orders to withdraw," Davis continued. "The order came just as the last Japanese assault began to sweep the survivors off the ridge. Those who could scrambled down the slope as fast as they could. . . . [O]nly nine made it down safely." Among the wounded was the captain, who had been hit in the legs by shrapnel.[19]
>
> – *Naval History Magazine,* October 2019

Company C's heroic fight for Hill 100 marked the end of major combat for Lt.Col. Davis' command, and on D+7 the badly depleted battalion went into landing-force reserve, having suffered 71% casualties.[20]

> General Smith reported . . . : "In this six-day assault against the high ground north of the airfield the 1st Marines engaged in one of the bitterest fights in the Pacific War. . . . [T]he regiment had suffered 1,737 casualties (209 killed in action and 1,438 wounded in action). It had killed an estimate 3,700 Japanese in the eight days it had been in the lines."[21]
>
> – *Naval History Magazine,* October 2019

Meanwhile, Just to the West of 2/1's Hill 200

2/7 had continued to climb the center portion of the southern Umurbrogol. The whole 7th Regiment was about to get more involved with the island's nerve center.

> On the 20th, the 1st Marines had to fall back from Walt Ridge to the southern side of an East Road causeway north of Asias. From this position these right-flank elements faced a 150-yard-wide valley bounded on the right by Walt Ridge and on the left partly by Five Sisters and partly by another ridge designated Five Brothers. This depression, known to the 1st Marine Division as Horseshoe Valley, was renamed Mortimer Valley by men of the 81st Infantry Division who later fought in the same area. In the center, during the 20th, elements of the 7th Marines pushed through strong defenses to high ground at the southwestern corner of a narrow defile which, lying west of Five Sisters, came to be

known as Death Valley. Left-flank units of the 1st Marines advanced up West Road a little farther, sending some troops up rising ground from the road to narrow the front of 7th Marines units in the center.

Despite some local advances, by evening on 20 September the 1st Marines had been stopped and,". . . as an assault unit on the regimental level, had ceased temporarily to exist" (Hough, "The Assault on Peleliu," p. 88), having suffered almost 1,750 casualties, well over half its strength. Relief was absolutely necessary, not only because casualties were heavy but also because the survivors were physically exhausted from heat, lack of water, and continuous combat. Elements of the 7th Marines therefore relieved all 1st Marines units except those along West Road.

During the next two days 7th Marines units in the center of the line could gain less than 175 yards; on the right Walt Ridge was again attacked to no avail; and on the left 1st Marines troops fighting along high ground above West Road suffered more casualties. Another attack against Five Sisters was forced back on the 22nd, and at dark containing lines were set up north of Asias, facing Walt Ridge and the mouth of Horseshoe Valley. After a week of extremely arduous combat over incredibly rough and well-defended terrain, the 1st Marine Division had been at least temporarily halted. The division had lost just under 4,000 men. The 1st Marines had been virtually eliminated from the fight (the 1st Battalion, for instance, was reduced to two understrength companies), the 7th Marines were little better off, and the 5th Marines as well as other units, especially Shore Party troops and combat engineers, had not escaped unscathed.[22]

— "Approach to the Philippines," by Smith
U.S. Army Center of Military History, 1996

The Re-Engagement of Hill 100

Though tired, 2/1 was not quite through. It had left Hill 100 under less than auspicious circumstances, and wanted to go back (probably to look for survivors).

> The attack on the morning of 0-plus 5 (20 September) had for its main objective the recapture of Walt Ridge [actually only Hill 100 at its end]. Once more 2/1 assaulted with the 1st Battalion (less Company A) and the Division Reconnaissance Company attached. Every resource under the regiment's control was employed. Every officer and man who could be spared from regimental and battalion headquarters was attached to the companies (1/1 Unit History, p. 28) Additional tanks, LVT(A)s, half-tracks, 37mm guns, and mortars were brought forward as far as possible. Deadly Japanese crossfire knocked out many, and those remaining proved inadequate to support the infantry successfully against the combination of cruel terrain and devilishly prepared positions.
>
> This was an all-out effort; there was nothing in reserve. Its result is described as well by the Japanese as in any Marine reports, and a great deal more succinctly:
>
> Since dawn the enemy has been concentrating their forces, vainly trying to approach Higashiyama (Walt Ridge) and Kansokuyama (Hill 300, at this time considered by Marines as part of the Five Sisters formation) with 14 tanks and one infantry battalion under powerful aid of air and artillery fire. However, they were again put to routoij, receiving heavy losses (Tada Record, 20Sept44).
>
> Although some advance was made, the low ground proved untenable, and the exhausted survivors fell back to a position where they could cover, at a respectful distance, the mouth of the Horseshoe. After six terrible days, it had to be admitted that the 1st Marines had been definitely stopped. More days were to elapse before graves registration men could get to the dead at the points of farthest penetration [not until 3 October].[23]
>
> — *The Assault on Peleliu,* by Hough
> History Division, HQMC, 1950

Company B had somehow helped with that second attack against Hill 100, possibly through an initial base of fire and follow-up assault.

> [Around] 19 September 1944. During an attack against a coral ridge when leading elements of the company were

> pinned down by an enemy machine gun which was strongly entrenched and well-concealed in a cave, Private First Class Manahan [machinegunner in B/1/1] observed the situation while firing a support mission and, without orders, displaced his gun forward and advanced in the face of heavy hostile mortar, machine-gun and rifle fire until he found an exposed position from which he could observe the Japanese automatic weapon. When his gun crew became casualties, Private First Class Manahan manned his gun alone and, although wounded four times, maintained a steady volume of fire at point-blank range until he silenced the enemy weapon.[24]
>
> — Navy Cross citation for PFC Merle E. Manahan

> 20 September 1944. Assuming command after his Commanding Officer had become a casualty during the initial fighting, First Lieutenant Rineer [of B/1/1] boldly led his company in repeated attacks against the enemy entrenched in dense jungle concealing a network of pillboxes. With only forty men left in his company on the morning of 20 September, he assaulted the steep coral slopes of Hill 100 in the face of intense fire emanating from hostile machine guns entrenched in coral caves and personally remained in the midst of the furious fighting to urge his men forward until seriously wounded in both legs. Steadfastly refusing evacuation, First Lieutenant Rineer courageously struggled forward to commanding ground and calmly organized his twenty-four remaining men into a defensive position from which they succeeded in accomplishing their mission.[25]
>
> — Navy Cross citation for 2ndLt. Francis Rineer

1st Marines Had Been Quite Near the Belly of the Beast

Over the full course of the battle, Col. Puller's 1st Marine Regiment would lose 1,749 of its 3,000 men (over 60%). General Roy Geiger then sent in U.S. Army elements to relieve the 1st Marines. The 321st Regimental Combat Team (RCT) landed on the west coast of Peleliu at the northern end of Umurbrogol mountain on September 23rd. Then, the 321st, along with the 5th and 7th Marines, took turns attacking the top of the mountain, all the while raising their casualty rates.[26]

Through mid-October, both the 5th and 7th Marines would lose approximately half of their effective fighting force. At this point, General Geiger decided to replace the entire 1st Marine Division with additional 81st Infantry Division soldiers. By the third week of October, the remaining Marines had been evacuated to Pavuvu.[27]

> The U.S. Army soldiers would continue fighting the Japanese forces . . . [atop the Umurbrogol] for another month before finally securing the island on November 27th ending 73 days of fighting. Prior to this, on November 24th, Nakagawa declared, "Our sword is broken and we have run out of spears." He then burned the regimental colors and performed ritual suicide. Anecdotally, a Japanese LT, 26 2nd Infantry [Regiment] soldiers, and 8 [eight] 45th Guard Force soldiers would hole up in the Peleliu caves until April 22nd, 1947, when a Japanese admiral convinced them that the war had been over for a while.[28]
>
> — *WorldWar2Facts.org*

How the Umurbrogol Pocket Would Ultimately Be Taken

Chesty's boys did their best to meet a nearly impossible time line. In the process, they had silenced many of the big guns targeting the flatlands.[29] But the southern Umurbrogol summit would be denied to the Americans for another two months. (See Map 8.2).

> The almost-sheer face . . . [atop Bloody Nose] is only a few hundred yards north of the airfield, [and] on the southern edge of the Umurbrogol complex. This particular coral peak and four others nearby were identified as the Five Sisters during the battle. The 1st Marines attacked Bloody Nose soon after landing on Peleliu and suffered devastating casualties in the failed attempt. Despite repeated attacks by Marine and Army units, the Japanese held this position for two months, almost to the end of their organized resistance.[30]
>
> — National Park Service Study I

American light planes were able to land on the airfield on

> September 19. That same day, the 1st Marines renewed the attack on the ridges, making small gains, but were again stopped at Bloody Nose Ridge. To the east, a company of the 7th Marines gained the crest of Walt Ridge but, suffering extremely heavy casualties, was forced to withdraw. During the next three days both the 1st and 7th RCTs assaulted the ridges time and again, without success. By September 22, the First Marine Division had suffered nearly 4,000 casualties, the 1st Marine Regiment alone losing 56 percent of its strength. General Rupertus decided to cease frontal attacks on the southern ridges; instead, he planned a drive up the west coast and an attack on the Umurbrogal [Pocket] complex from the north.[31]
>
> — National Park Service Study I

The U.S. Army's way of doing things relied more on direct bombardment than that of the Marines. Until the Umurbrogol's northern end could be climbed, the terrain would be too broken for this style of warfare. That's when the landing force reserve was summoned. After helping the 322nd to seize Angaur, the 321st RCT came ashore at Peleliu. Soon thereafter, the 323rd joined the fight for the mountaintop.

How Peleliu Would Be Remembered

The whole painful Peleliu experience soon faded into relative obscurity. Though that lengthy battle had been the Corps' most challenging to date, it rated only one short paragraph in the official heritage book of 1998.[32] Within the hardest-fought battles lie the most honor and insight. The U.S. Army hadn't rescued the Marines at Peleliu, just closed down the enemy's last-stand operation. That's not much different from its arrival at Guadalcanal and several other islands, so the Marines could go on to the next amphibious landing.

Now, most Leathernecks only realize many GIs got hurt trying to take an island of questionable value to the overall war effort. Such things occur all the time in combat because proficient adversaries do the unexpected.

Though Peleliu became less necessary to the capture of the Philippines, it and Biak still protected the main U.S. supply lines in the region, as well as hard-won New Guinea. Within the well-fortified Palaus had existed a considerable air threat—several good airfields and foolproof ways to protect their planes from bombardment. Besides easy to excavate limestone cliffs, Peleliu had the cement covers to now missing mounds of dirt. Such hangars were nearly invisible from overhead. And the Palaus constituted Japan's principal resupply and reinforcement base for everything south of there. Rabaul and Truk were far to the east.

So, Peleliu had retained much of its strategic importance, and its seizure not involved any major embarrassment for America. Those in charge of the ground operation did a good job—Gen. Rupertus included. He didn't invent the U.S. way of war. The underlying problem had been the worst terrain anywhere in the Pacific (with the possible exception of Biak). Peleliu's protruding coral had blocked pre-assault bombardment and access to the Japanese control center.

What happened at Peleliu should be remembered as one of the most glorious battles in USMC history. The 1st Marines—Chesty Puller's outfit—had fought, and fought, and fought. Against a highly deceptive and well-dug-in foe, that takes a little bleeding. Much of its final effort on Peleliu was directed toward the south end of Walt Ridge. That's less than 200 yards from the original enemy CP. Enemy Reserves were likely reinforcing Hill 100 from below ground.

So, the fighting spirit of the 1st Marine Regiment, then the 7th and 5th, and finally the 81st Army Division are forever inscribed on the walls of honor. Their dogged determination and tactical countermoves would inspire a winning effort on Okinawa.

11 Degree of Tunneling in the Southern Massif

- Could tunnels have connected many enemy positions?
- How often was the foe seen maneuvering above ground?

Some of the ground had to be continually retaken.

(Source: "Bloodly Beaches: The Marines at Peleliu," by B.Gen. G.D. Gayle, Marines in WWII Commemorative Series, p. 48.)

The Rest of the Story

The 1st Marines had advanced too quickly on Peleliu to seal every cavity they found on the forward slopes of the Umurbrogol, much less look inside. As a result, neither the official histories nor award citations talk much about cave interiors. But, Peleliu was the Corps' first exposure to an extensive sub-surface realm. There had been artillery in caves before, but not to extent that Peleliu was laced with underground chambers. Some had been naturally created and others dug. Yet, the invasion force was practically blind to their infiltration potential—to the unobstructed access they provided to U.S. command and logistics networks. Each time a Japanese cave or pillbox was overrun, it became the potential source of rear-area activity. (See Map 11.1.)

2GW's Subliminal Assumption

2GW is a warfare style for commanders sure their supporting arms will eradicate most opposition—that their quarry will be unable to resist a ground assault. Because the enemy's countermeasures no longer matter, the 2GW proponent makes too little provision for the foe's thought processes.

If the Japanese defenders of Peleliu had been the least bit alert, they would have carefully safeguarded their underground system. Four things would have occurred at every tunnel entrance: (1) sniper security; (2) boobytrapping; (3) the preemptive collapse of any passageway within a jeopardized cave; and (4) a camouflaged entrance to all interior tunnels. So, no one should be surprised if such things happened on Peleliu. Every entrance to the huge 1000-man cave at the north end of the island would take four months to be discovered.[1]

Tunnel-Protecting Snipers

Vietnam War veterans well remember the sniper fire around any portal to a major underground facility of the Viet Cong (VC). Some Peleliu reports indicate a similar practice.

> A small natural cave [AB207] exists in the side of a narrow pass between the limestone cliffs and has been collapsed. According to our guides, this is the unmarked site where [the much beloved] Captain Haldane was killed by a sniper as described by Eugene Sledge.[2]
>
> — National Park Service Study II

> AB210. Sniper Position. Located about 20m north of . . . the [Navy] . . . Avenger wing is a small rock shelter facing NE, measuring 2m wide, 2m deep and 60cm high. Several spent rifle cartridges remain on the surface and it is

> a possible firing location for the sniper that killed Captain Haldane (site AB207).[3]
>
> — National Park Service Study II

Boobytrapped Tunnel Entrances

U.S. troops seldom inspected the caves they encountered on Peleliu for a reason. Any sub-surface space was risky to be around. Besides die-hard occupants, it often contained concealed explosives that had been specially rigged to discourage trespassers.[4]

> More than 118 caves were among the 285 WWII sites documented on Peleliu in just nine days of fieldwork. . . . Exceptionally dangerous sites were addressed immediately by the UXO team, such as booby-trapped caves. One cave contained 27 land mines each placed on a grenade, while another presented us with a carpet of 97 grenades with all the pins pulled.[5]
>
> — National Park Service Study II

Best protected would have been the tunnels that led to stores of food, water, and ammunition.

> On the east side of the room [in Japanese storage cave AB160], on the wall opposite the entry, there is a 5m wide break in the wall which drops about 3 m down into another passage.[6]
>
> — National Park Service Study II

> A Japanese storage cave [AB160] . . . is one of the WWII sites on Peleliu that also has traditional cultural significance for Palauans. . . . This cave may well be the second such storage cave that Denfeld had heard about but was unable to locate (Denfeld, 1981:70). The concrete lined entry faces west and has been sealed with rubble since 1944, however a large shell hole above the doorway now serves as a new entry [point]. . . . The cave contains . . . a Japanese 50 Kg. bomb, . . . numerous live Japanese land mines with three exposed on the floor and others probably hidden beneath the bat guano which is thick on the floor. The presence of

> UXO [unexploded ordnance] and the well known location and easy accessibility of this cave combine to make this a dangerous site until the UXO can be removed.[7]
> — National Park Service Study II

Closing the Mouth of Internal Passageways

Most communication tunnel entrances on Peleliu were inside naturally occurring caves. The National Park Service later discovered quite a few caves that had been partially destroyed from within: e.g., AB87, AB100, AB108, AB116, AB129,[8]

> Some features [within Peleliu's central ridge system] represent rooms and passages remaining from an ancient collapsed cave systems.[9]
> — National Park Service Study II

Camouflaged or Unlikely Tunnel Entrances

Not until Vietnam would GIs learn the most clever ways of entering an underground facility. Its lessons included: (1) within the wall of a well; (2) below the waterline of a deep creek; (3) on the side of a rice paddy dike; (4) under a man-made structure.

On Peleliu, there were wells near the rectangular blockhouse and runway boundary that had spawned big counterattacks. (Look back at Maps 7.1 and 7.4.) A huge open-pit gravel mine existed at the end of Walt Ridge. (Look back at Map 10.1.) And not just the Momiji Plateau had been covered with sink holes. Within Chapter 7 was one at the end of a flatland peninsula.

> Instead of a gently rounded hill, the [whole] Umurbrogol area was in fact a complex system of sharply uplifted coral ridges, knobs, valleys, and sink holes.[10]
> — *Bloody Beaches,* by Gayle,
> Marine Corps History Center, 1996

An open lagoon-front cave had been near the circular blockhouse where a counterattack happened. And there may have been others with underwater access.

Then, there were the scores of emplacements mysteriously re-manned during the battle. The following position could clearly be reached by tunnel.

> This Japanese defensive feature [AB6] consists of a collapsed dugout that once had a wooden plank roof and a low coral rock revetment. The south end of the dugout adjoins a low natural cave no more than 1 - 1.5m high. Two narrow tunnels branch out from the far corners of the natural cave. These were deemed too unsafe to explore further.[11]
> — National Park Service Study II

While no trapdoor was ever reported in the Palaus, they almost certainly existed. The 1000-man cave that took four months to explore had more than one level. So, fully imagining Peleliu's underground network takes only some knowledge of other Asian wars. But, the most abundant tunnel entrance on Peleliu had probably been the natural crevices too small to admit the average GI. The next chapter will discuss a crack in the rock from where 30-50 Japanese guerrillas successfully operated until April 1947.[12]

Number of Caves on the Umurbrogol's Forward Face

Of the 500 caves found on Peleliu after the battle, Map 7.5 shows only about 50 at the southern end of the Umurbrogol. A closer looks at the contour lines on Map 11.2 reveal most to have been on the mountain side above the airport. That's because most in the Pocket were eradicated in 1944. And 1/1 alone had reported 35 on its first day up the hill.

Map 11.2. only depicts 17 caves in places the Marines had initially come. Two or three were where 2/7 climbed Hills 180 and 160. Several more existed between Death Valley and the ridgeline formed by Hills 200 and 205. Of course, many more had once existed. The Park Service would subsequently investigate 118 in or around the Momiji Plateau.[13]

But, Map 11.2 also reveals caves on the southeastern flank of the mountain. Four were astride Five Brothers overlooking Hill 100 on Walt Ridge. And four more were at the base of that Hill 100. So many in a row may indicate the finger fortified from within. Most amazing were the five at the mouth of Horseshoe Valley.

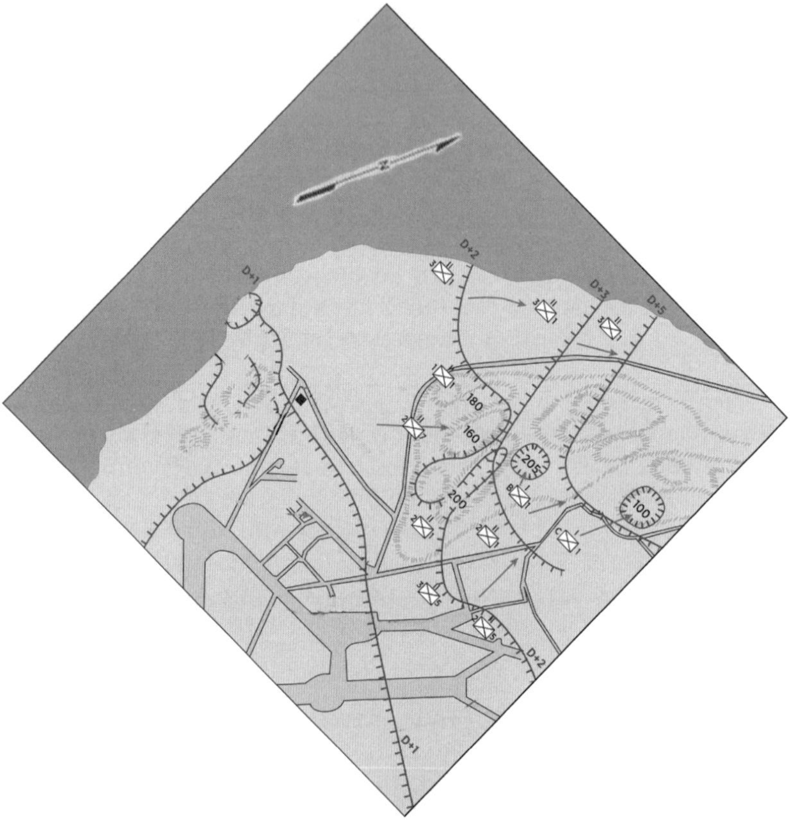

Map 11.1: Initial Marine Advances Up the Umurbrogol
(Source: Courtesy of "Naval History Magazine," from "Eight Days in Hell," October 2019 issue, © 2019 U.S. Naval Institute/www.usni.org.)

They were no more than 10 meters above sea level. As close as these caves had been to the original Japanese headquarters, it's small wonder the 1st Marine Regiment had so much trouble in this area.

Tunnels Associated with Defense Line Cracks in the Massif

The caves within the steep sides of the defense line crevasses

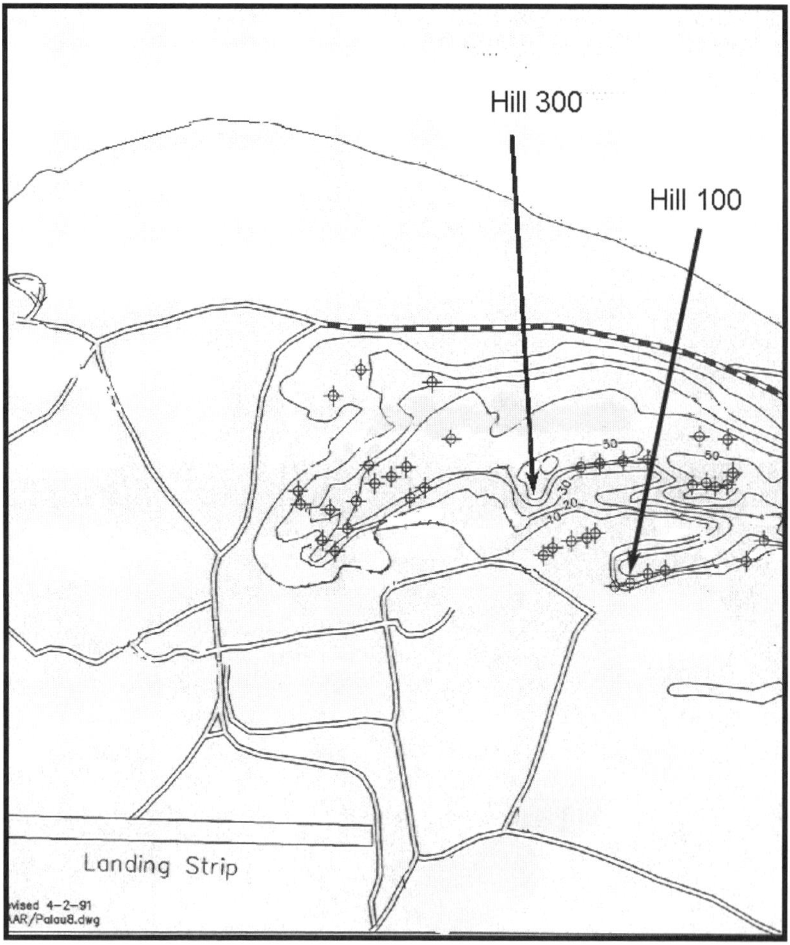

Map 11.2: Frontal-Face Caves on the Umurbrogol
(Source: "Preliminary Draft -- 1/91 (of Peleliu study)," Botany Dept., University of Hawaii, January 1991, fig. 9.)

were hard to reach and practically immune to standoff firepower. (Look back at Figure 7.2.) Their occupants guarded the bottom of each defile with overlapping sectors of fire.

> Every advance opened the advancing Marines to new fire from heretofore hidden positions on flanks, in rear, in caves above or below newly won ground.[14]
>
> — *Bloody Beaches*, by Gayle,
> Marine Corps History Center, 1996

At this nightmarish barrier, enemy explosives could come flying in on the Marines from any direction. Those from above were harder to counter.

> [25 September] When a Japanese soldier emerged from a cave in a cliff directly below an observation post and suddenly hurled a grenade into the position from which two of our men were directing mortar fire against enemy emplacements, PFC New [of 2/7] instantly perceived the dire peril to the other Marines and . . . flung himself upon the grenade.[15]
>
> —Medal of Honor citation for PFC John Dury New

Those unexpected grenades had made it virtually impossible for the young Leathernecks to cross several deep trenches in succession.

> Shortly after his leader ordered a withdrawal upon discovering that the squad was partly cut off from their company . . . during an aggressive attack on the strongly entrenched enemy, PFC. Roan [of 2/7] and his companions were suddenly engaged in a furious exchange of hand grenades by Japanese forces emplaced in a cave on higher ground and to the rear of the squad. Seeking protection with four other Marines in a depression in the rocky, broken terrain, PFC. Roan was wounded by an enemy grenade which fell close to their position and, . . . when another grenade landed in the midst of the group, unhesitatingly flung himself upon it.[16]
>
> — MOH citation for PFC Charles Howard Roan

To further complicate matters, all those Japanese crevasse sentries were somehow linked to the Umurbrogol's central tunnel system. Their reinforcements must have come through a backside entrance to their lofty perch. To withdraw, they had only to follow a well-rehearsed path at the bottom of the subsequent declivity. Somewhere along that path was a tiny crevice that doubled as tunnel entrance.

> 3 October 1944. After his company had occupied a narrow ridge and the safety of its positions was jeopardized by heavy enemy machine-gun fire from positions which completely commanded the ridge, Private First Class Pottersnak [of 3/7] voluntarily crossed over the ridge and, extricating a hostile machine gun from a cave in which its crew had already been killed, destroyed the enemy weapon before returning to his own lines. Observing that a group of the Japanese was attempting to put another machine gun in position [at the same location], he again crossed the ridge and destroyed the enemy gun and crew.[17]
>
> — Navy Cross citation for PFC C.J. Pottersnak

PFC Pottersnak probably did most of his cave clearing with rifle grenades, but a bazooka would have also been good for this type of work.

The Walt Ridge Tunnel Network

When Hill 100 was first assaulted on 19 September by C/1/1, "[it was] a heavily defended coral nose that had been studded with caves and concrete pillboxes." The caves had also been fortified, because PFC Manahan suppressed heavy fire coming from a Hill 100 cave.[18] (See Maps 11.3 and 11.4.)

When B/1/1 assaulted Hill 100 again on 20 September, it also encountered "intense fire emanating from hostile machine guns entrenched in coral caves."[19] Most Japanese gun sites were mutually supporting. In other words, the forwardmost site could not be captured without concurrently engaging those behind it.

> 20 September 1944. After his company had been cut down by hostile machine-gun fire during an assault over open ground, Private First Class Harris [of Hdqts. Co./1/1] boldly returned to the area through intense fire from automatic weapons emplaced in caves on the crest of a ridge and, sin-

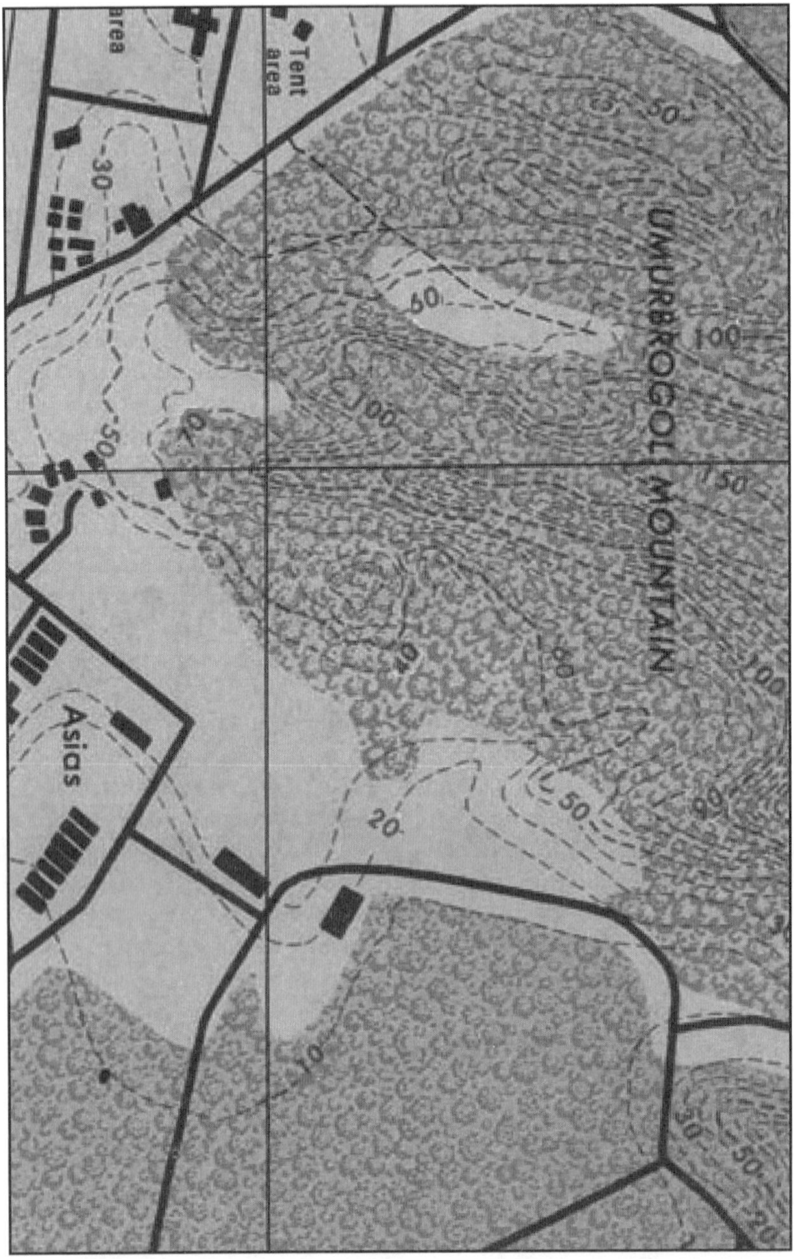

Map 11.3: Topographical Map of Southern Umurbrogol
(Source: Perry Castañeda Collection, Univ. of Texas at Austin, Palau Islands 1:12,500, Series W854, U.S. Army Map Service, 1946, Ngarmoked NW-D, Sheet 7331 I NW.)

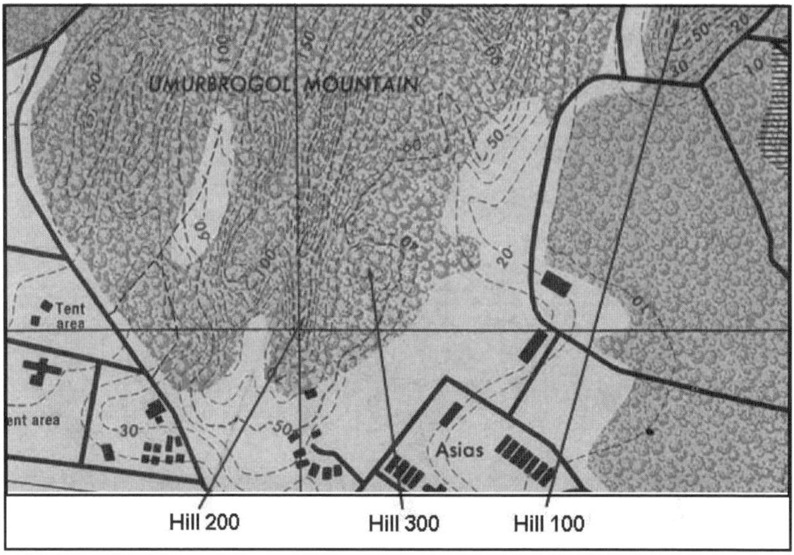

Map 11.4: Key Locations Within the Many Contour Lines
(Source: Perry Castaneda Collection, Univ. of Texas at Austin, Palau Islands 1:12,500, Series W854, U.S. Army Map Service, 1946, Ngarmoked NW-D, Sheet 7331 I NW.)

> gle-handedly turning on the first cave, emptied his pistol into the enemy gun crew. After silencing the weapon with hand grenades, he picked up an automatic rifle and, delivering intense fire into a second Japanese position, enabled two men of his company to crawl to safety.[20]
>
> — Navy Cross citation for PFC Boone T. Harris

That the Hill 100 Marines kept getting counterattacked may indicate a tunnel leading all the way back to Nakagawa's Reserves. (Look back at Map 7.1.)

Far Less Tunneling Within the Hill 200 Ridgeline

Hill 200 was at the southern tip of a tall finger that had paralleled East Road. (Look again at Maps 10.3 and 10.4.) Map 11.2 shows several caves within that grouping of big Umurbrogol fingers.

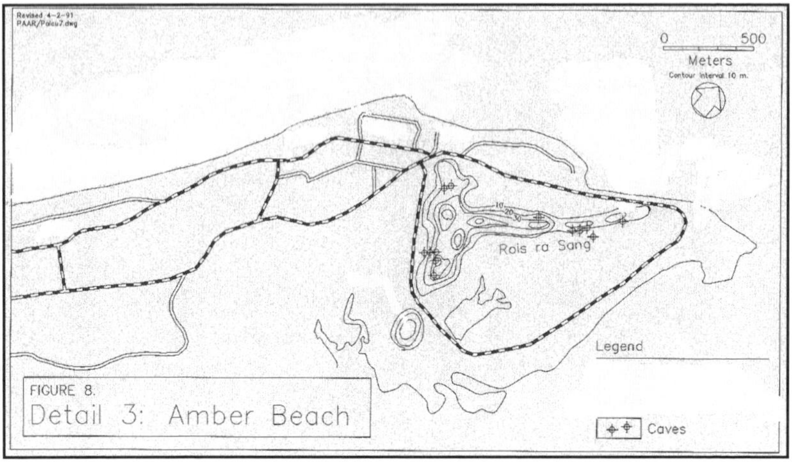

Map 11.5: Caves in Amiangal Mountain

(Source: "Preliminary Draft -- 1/91 (of Peleliu study)," Botany Dept., University of Hawaii, January 1991, fig. 8.)

2/1's assault on Hill 200 revealed the ridgeline's degree of fortification. Around 18 September, PFC Robert L. Bungard of G/2/1 destroyed "four hostile pillboxes and cave entrances."[21] Among those enemy positions had likely been the lone cave that Map 11.2. shows at the base of Hill 200.

After reaching the top of Hill 200, Corporal Kincaid of E/2/1 volunteered "to occupy a sharp crest which afforded the only protection to his company's flank position on a ridge." Late that night, an enemy mortar crew emerged from a cave [higher up the finger] and attempted to set up a heavy weapon.[22] So, unlike Walt Ridge, the crest of this Hill 200 ridgeline was not swarming with defenders. Yet, during the initial assault, 2/1 had still faced "point-blank salvos from mountain guns . . . which were run in and out of cover."[23] So, at least one hillside tunnel likely supported those guns.

The Umurbrogol Pocket Passageways

During the pre-invasion bombardment, enemy headquarters

had been deep below Old Baldy, according to Map 4.3. Then, that CP was moved to the Momiji Plateau. It and most of the Umurbrogol Pocket emplacements were likely connected below ground. Its evidence is in extent of the new headquarters excavation and natural hollowness of the Plateau.

> This cave [A81] was among those documented by the military as part of the post-battle study of the Japanese caves on Peleliu. At that time, the cave penetrated the ridge between Death Valley and Wildcat Bowl, but those passages have since collapsed or been sealed. According to Phelan (1945:5) this cave is unusually significant in that it was used as the Japanese command center after a retreat from the Japanese Navy Command cave, a large natural cave in the southern ridges of the Omleblochel (site AB146).[24]
> — National Park Service Study II,

Peleliu's Other Big Underground Network

Amiangal Mountain at the north end of Peleliu was also filled with subterranean chambers and passageways built by a Japanese Naval Construction unit. (See Map 11.5.) Many housed heavy weapons. It would be the 5th Marines' job to silence those weapons. Sometime between 15 and 29 September, with his squad pinned down by hostile automatic weapons fire from the mouth of a cave dug into the side of a strategic hill, Cpl. Nicholas of B/1/5 "tirelessly directed his squad to fire point-blank into the cave's entrance so that tanks and flame throwers could finish the job."[25] Of course, Leatherneck engineers would eventually be needed to subdue this multi-tiered bastion.

> 4 October 1944. Vigorously supporting the advance of a friendly company, Sergeant Sheehan [of 1st Engineer Bn.] led his demolition squad with furious aggressiveness. When the enemy hurled the full force of their huge gun-batteries against the onrushing Marines and held them up with the merciless enfilade fire of one cave-emplaced machine gun which struck down forty-two of the forty-six men in the advancing unit and forced a temporary withdrawal, he fearlessly pushed on alone despite the ceaseless pound-

ing of the heavy Japanese weapons. With the hostile cave as his objective, he relentlessly worked his way forward and, reaching his goal, dropped a powerful demolition charge into the Japanese position, destroying the valuable machine gun and annihilating the entire gun crew.[26]

—Navy Cross citation for Sgt. Joseph D. Sheehan

12 Foe Who Kept Popping Up Behind U.S. Lines

- How many enemy soldiers popped up behind U.S. units?
- How did they get there?

U.S. artillery could only fire to the front of GI positions.

(Source: "Bloody Beaches: The Marines at Peleliu," by B.Gen. G.D. Gayle, Marines in WWII Commemorative Series, p. 47.)

Strange Things Were Still Happening

By September 1944, Japanese soldiers had become well known for their short-range infiltration abilities. So too had U.S. Marines become good at stopping intruders from crawling through their lines. So, when enemy soldiers kept showing up where they shouldn't have been on Peleliu, only a few U.S. leaders guessed why.

> 15 September 1944. Before permitting his men to use an enemy dugout as a position for an 81mm mortar observation post, 1stLt. Rouh [of 1/5] made a personal reconnaissance

> of the pillbox and, upon entering, was severely wounded by Japanese rifle fire from within. Emerging from the dugout, he was immediately assisted by 2 Marines to a less exposed area but, while receiving first aid, was further endangered by an enemy grenade which was thrown into their midst.[1]
>
> — Medal of Honor citation for 1stLt. Carlton Rouh

Shortly after the landing, two U.S. engineers assaulted an emplacement that had been "reoccupied by infiltrators." The pair of Marines then destroyed the emplacement and its eight new occupants.[2] As engineers, they may have suspected a sub-surface entrance.

U.S. Leaders Seemed Unaware of Peleliu's Stranger Aspects

Somebody had failed to apprise American infantry officers of the geological history of Peleliu's central hill-mass.

> The same [ocean bottom] pressure . . . [had] created innumerable underground faults in the coral limestone; . . . many of which had been further enlarged by [rainwater] erosion.[3]
>
> — *The Assault on Peleliu,* by Hough
> Historical Division, HQMC, 1950

Nor had those young Yank leaders heard of the Japanese tunneling on Peleliu. In combination, these two activities—one natural and the other made-made—could have fully laced that tiny island with caves and sub-surface passageways. Such a secretive aspect to any battle space would give its defenders an edge in maneuver. Not until half-way through the struggle for Iwo Jima in March of 1945 would any U.S. commanding officer (CO) report an adversary traveling any distance below ground. Then, the particulars of the discovery were astounding. After every engagement, strongpoint occupants had been moving below ground to rearward strongpoints.[4]

But on Peleliu, none of the chronicles ever suggested such a thing. All unusual enemy activity was attributed to above-ground infiltration. And the culprit ended up in a previously occupied bunker because it was the only available cover. Those in charge of the operation only admitted to Japanese soldiers occasionally hiding in place to let the Americans pass overhead.[5] Yet, the possibility of "passive infiltration" was so remote that most COs ignored it. Then, pouring diesel fuel down every opening and sealing it with explosives seemed the perfect solution. The assumption was that every sub-surface inhabitant would die of suffocation. But, where an underground labyrinth of any size was involved, he didn't. Even when completely cutoff from the surface, Japanese cellar-dwellers simply excavated a new exit.

> It would take another several weeks to finally eliminate all resistance on the [Akarakoro] Point, by blasting shut the cave entrances. Even then, a few weeks later, the Marines were astonished to see Japanese Navy survivors dig their way out! [6]
>
> — "Battle of Peleliu," *HistoryofWar.org*

Most Western society members find the whole idea of three-dimensional warfare counter-intuitive. They don't look for people in trees or beneath the soil. Early in the invasion of Peleliu, tunnels were discovered once—at the big blockhouse in the 1st Marines' sector. They connected satellite bunkers to a central bomb shelter. No underground extension to the Umurbrogol was ever suspected. Not until the final assault on the Southern pocket would sub-surface passageways be credited for the constant harassment by "infiltrators."[7]

Of course, there's another explanation for so many unexpected visitors in the Palau Islands. Many of the Japanese strongpoints may have had secret rooms beneath them, and a hidden trapdoor. This kind of trick would be commonplace in Korea some six years later.

> Whether we can defend our positions is a question that was resolved last year. The solution was to hide in grottos *[sic]*. We dug out a two-level fortification. . . . Once the enemy entered the surface positions, we started counterattacks and inflicted heavy casualties on them [from below]. By this crude means, we were able to seize and take away the equipment left by the enemy.[8]
>
> — *Mao's Generals Remember Korea*

Along Roadside Cliffs Arose Other Problems

Once the southern part of the island had fallen into U.S. hands, its two northbound roads offered a less dangerous way to reach the summit of the Umurbrogol. But, along the West Road lay some imposing cliffs and an abundance of infiltrators.

> The only tactical option along the West Road was to seize and hold the coral spires and cliffs commanding the road, and to defend such positions against infiltrators. Once those heights were seized, troop units and trucks could move along West Road. Until seized, the "cliff" offered concealment and some cover to occupying Japanese. Until those cliff positions were seized and held, the Japanese therein could be only temporarily silenced by heavy firepower. Until they were driven from their commanding positions, the road could not be treated as truly open.
>
> Those terrain conditions existed for three-quarters of a mile along the West Road. There, abreast [of'] the north end of the Pocket, the plateau of coral sinkholes merged into a more systematic group of limestone ridges.[9]
>
> — *Bloody Beaches,* by Gayle,
> Marine Corps History Center, 1996

There was another terrain-associated problem some 500 yards beyond where the West Road started north along the hill-mass. Between road and coast was a sub-surface facility in which 30-50 Japanese guerrillas would hold out until April 1947. Its tiny entrance lay at the base of the mountain between squat Umurbrogol fingers that ran under the road. No one has since crawled through the crevice at the front of a miniscule cave.[10] That cave must have contained a further passageway to a bigger space—possibly the mountain's central tunnel network.

> The holdout cave is located about 100 m[eters] east of the West Road and is adjacent to a large mangrove swamp and marsh filled with standing water. The entrance to the natural cave faces west and is a narrow vertical slit 1.05m long and 35cm wide, small enough to be easily hidden under a few palm fronds during the day. Small arms rounds can

> be seen inside on the floor of the cave which is just over 1m high.[11]
>
> – National Park Service Study II

While Moving Farther North

As 7th Marines worked along the high ground, 1/321st seized the bottom part of West Road. But then, other incidents started to occur along that road that were difficult to explain.

> The 3rd Battalion [of 321st Infantry] advanced northward over the same ground traversed by 1/321 two days previously, combing the area for any Japanese who might have been missed by its predecessors. A number of such were found and put out of the way (81st Opn.Rpt.H, 31). Proceeding with that systematic thoroughness characteristic of all 81st Division units on Peleliu, the battalion paused to blow in every cave the men were able to discover, as a precautionary measure against reoccupation by infiltrating Japanese.[12]
>
> – *The Assault on Peleliu,* by Hough
> Historical Division, HQMC, 1950

The Japanese strategy was for the first time clear. While there had been no Biak-like *Haichi Shiki*, several groups of infiltrators were making the road too dangerous to travel.

> Here is another example of Japanese allowing themselves to be overrun in order to get in American rear, since the 1st Battalion had traversed this same ground two days earlier without encountering any of the enemy.[13]
>
> – *The Assault on Peleliu,* by Hough
> Historical Division, HQMC, 1950

On the way up to where the West Road merged with its eastern counterpart, the assault force ran into more enemy "replacements."

> After a night marked by the killing of about 25 Japanese

> infiltrators in its sector, the 1st Battalion, 321st Infantry, resumed its attack north[ward] on the morning of 28 September. . .
>
> Meanwhile, Company C had found it necessary to fight its way back up Hill 80 [in the saddle between northern and central ridge systems], which Company I, 5th Marines, had left late the previous afternoon.[14]
>
> — "Approach to the Philippines," by Smith
> U.S. Army Center of Military History, 1996

Flatland Counterattacks Out of Nowhere

Chapter 7 discusses how counterattacks were launched from Peleliu's two biggest blockhouses. Map 7.4 suggests the northern one came through a tunnel from the Umurbrogol. That same map shows an assault force suddenly appearing next to an airport runway. According to Map 7.1, there were wells at both locations. If the Japanese had been using wells as exits to a flatland tunnel system, foe Reserves from beneath the Umurbrogol could have conducted both attacks.

Infiltrator Activity Just Outside the Pocket Cordon

As U.S. forces attempted to empty the Umurbrogol Pocket, the number of rear-area setbacks increased along the principal access routes to it.

> On 24 November, Colonel Nakagawa . . . said that the final 56 men had been split into 17 infiltration parties, to slip through [or under] the American positions and to "attack the enemy everywhere."[15]
>
> — *Bloody Beaches,* by Gayle,
> Marine Corps History Center, 1996

Between ridgelines defining the Momiji Plateau was the natural equivalent to an underground parking garage. That's how plateau defenders moved around so readily.

WWII era descriptions of the interior of Peleliu describe

> the complex topography that the Japanese defenders employed to their advantage. The maze of steep walled bowls and sinkholes are so complex that even today they defy our ability to map them. Some features represent rooms and passages remaining from an ancient collapsed cave systems and combat in this area was akin to fighting in an urban city, one that was virtually bombproof and that had to be taken room by room.[16]
>
> — National Park Service Study II

After such intense enemy preparation of the battle space, the fighting turned chaotic and costly.

> The next morning [23 October] the two [323rd Infantry] rifle companies pushed most of the remaining Japanese north toward Death Valley and China Wall, but final mopping up at the Southern Pocket was not completed until 3 November, after the line of attack had been changed from north to south. Altogether, over 100 Japanese were killed in the Southern Pocket, at the cost of 22 Army and Marine Corps troops killed and 51 wounded. While the action had no significant effect upon operations farther north, it is indicative of the Japanese ability to infiltrate over (and literally through) the Umurbrogol's broken terrain and to hold reoccupied positions against great odds.[17]
>
> — "Approach to the Philippines," by Smith
> U.S. Army Center of Military History, 1996

This Japanese counteroffensive was substantial enough to suggest the reinforcement of forces still resisting well beyond the Pocket.

> The Japanese still strong enough to act vigorously to the 321st Infantry's advances, attempted to infiltrate through the lines Company I had set up on the floor of Mortimer [a.k.a. Horseshoe] Valley, probably in an effort to reoccupy caves on Walt Ridge's western base.[18]
>
> — "Approach to the Philippines," by Smith
> U.S. Army Center of Military History, 1996

13 Initial Access to Top of the Umurbrogol

- What hills had to be taken to ascend the Momiji Plateau?
- How much equipment was brought up to that plateau?

321st gets first crack at north end of Momiji Plateau.

(Source: "Bloody Beaches: The Marines at Peleliu," by B.Gen. G.D. Gayle, Marines in WWII Commemorative Series, from public domain cover art by Tom Lea.)

More Non-Assailable Terrain

At the top of the Bloody Nose Ridge lay a moat-like canyon. The southwestern corner of the Umurbrogal was ringed by parallel cracks in the earth. And, most of the mountain's northwestern side sported a summit cliff. Along all three terrain features, the 7th Marines would establish their portion of the cordon around the high plateau. (See Map 13.1.) Only 2/7 had prior experience with those cracks.

When the 81st Infantry Division entered the picture, the sister Services took turns chipping away at the Pocket.

> [T]he corps commander, General Geiger, had to step in and direct the 321st Infantry Regiment to move from Angaur to Peleliu, where it landed September 23. The depleted 1st Marine Regiment was taken off Peleliu on September 30, its casualty figure having reached 1,672. . . .
>
> On September 24, [Gen.] Rupertus sent his . . . [newly attached] soldiers on a push up the West Road. The 321st RCT reached its objective, discovering along the way a trail running across the ridges toward East Road. A company, directed to explore the trail, moved eastward and succeeded in capturing Hill 100, the northern bastion of the Japanese defenses in the Umurbrogal ridges.[1]
>
> — National Park Service Study I

This was a different Hill 100 from the one 1st Marines had tackled on Walt Ridge. (Look again at Map 13.1.) But, before the Momiji Plateau could be crossed from north to south, the 5th and 7th Marines had to move inside the MLR that encircled the Umurbrogol summit. Some of that effort would be at the heavily fortified Five Sisters. Other engagements occurred at the northern end of Walt Ridge (to include Boyd Ridge). (See Map 13.2.) None were easy.

> The 7th Marines continued to press from the south and west while . . . Hill B, which had stalled [the] 321st RCT's attack, was finally taken on D+11 [26 September 1944] by 2/321. . . . 1/7 and 3/7 relieved 321st RCT on D+14 and on D+15 renewed the assault southwards, managing to take part of "Boyd Ridge" and Hill 100. . . . [Then] 3/5 . . . reinforced the 7th Marines on D+18, and . . . the regiment planned a four-battalion attack: 1/7 (along the East Road towards the unnamed ridge), and 3/7 (towards Baldy Hill) would attack from the north. 2/7 would attack towards Hill 300 from the south and 3/5 would make a diversionary attack towards Five Sisters . . . from the west. After bitter fighting and heavy casualties, the assault managed to secure its objectives with the exception of the Five Sisters, where 3/5 had managed to scale four out of the five heights but had to retreat as its [new] position was untenable. . . .
>
> The 7th Marines had been in the Umurbrogol for two weeks and were looking severely battered as a result. D+19

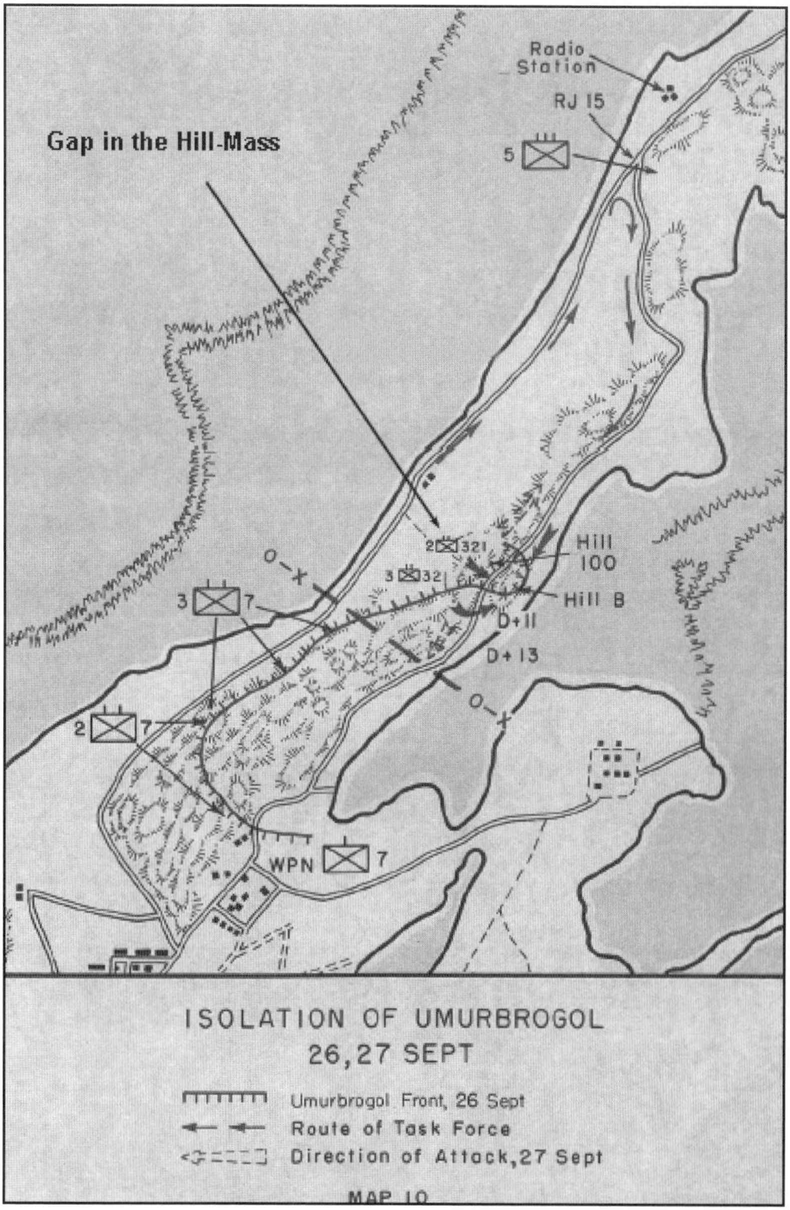

Map 13.1: 321st Infantry Followed a Trail Through the Hill-Mass
(Source: "The Assault on Peleliu," by Maj. Frank R. Hough USMCR [Washington, D.C. Hist. Div., HQMC, 1950], map 10.)

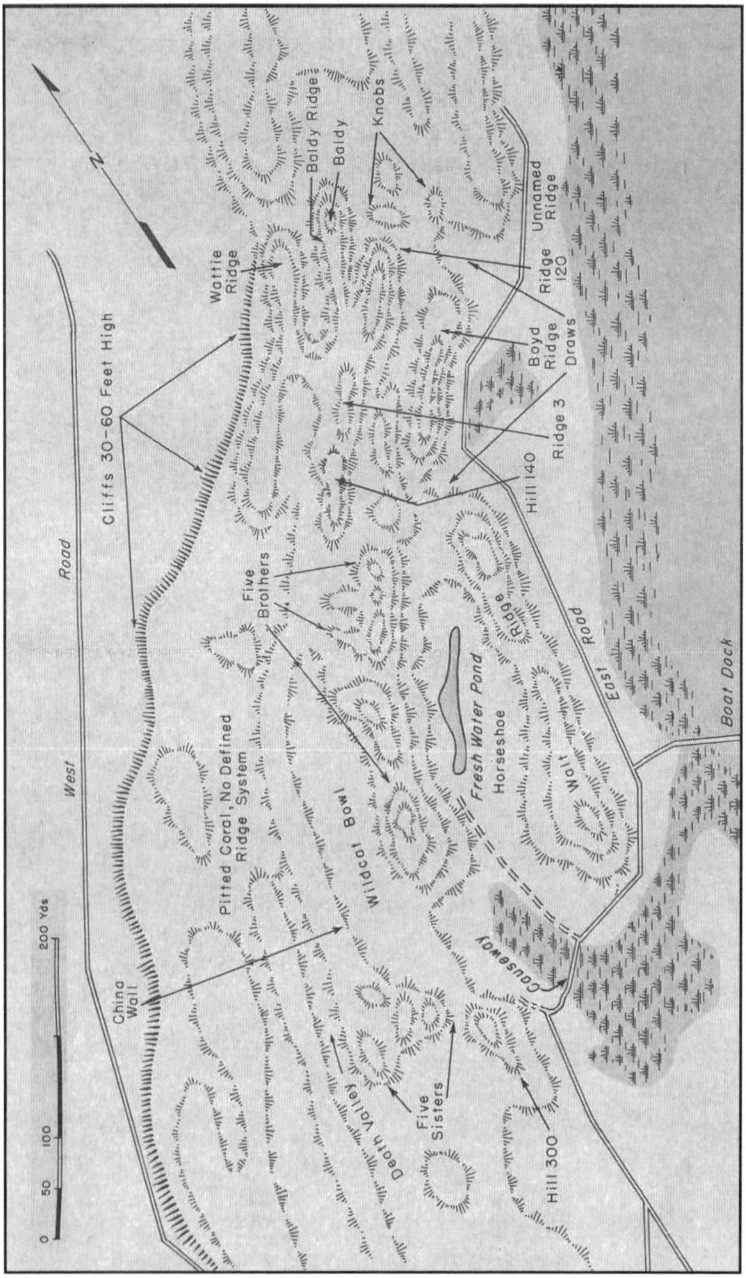

Map 13.2: Rest of the Fight for the Umurbrogol Pocket
(Source: "The Assault on Peleliu," by Maj. Frank R. Hough USMCR [Washington, D.C. Hist. Div., HQMC, 1950], map 14.)

saw their final attack get underway to mop up the draw between Walt (also known as Hill 100) and Boyd Ridges.[2]
— "Battle of Peleliu," *HistoryofWar.org*

Overall Success of That Final Marine Push

In this effort to shrink the Pocket, the Marines would capture several major objectives from the die-hard Japanese. (See Map 13.3.)

> On September 30, the 7th Marines began an assault on the Umurbrogal ridges in an effort to reduce the pocket, which measured about 1,500 yards north to south and 500 yards in width. Elements of the 5th Marines joined the attack the following day. By October 3, the Marines had captured the crests of two important ridges, Walt and Boyd, on the eastern side of the pocket, thus opening East Road. The 7th Marines, particularly, experienced heavy casualties in this four-day fight. By the evening of October 4, the 7th "was no longer able to function as an effective combat unit on the regimental level" (Garand and Strobridge, "Western Pacific Operations," 236). Rupertus ordered the 5th Marines to relieve the 7th.
>
> Between October 6 and 14, the 5th Marines made repeated attacks on the pocket, steadily making small gains, particularly in the north and northwest. It captured such landmarks as Baldy Ridge, Ridge 120, Ridge 3, Hill 140, and knobs of coral not named. Then, on October 14, General Rupertus withdrew all the First Marine Division from the battle lines, replacing it with the 321st RCT and a battalion of the 323rd Infantry Regiment freshly arrived from Ulithi Atoll.[3] [See Map 13.2.]
> — National Park Service Study I

Specifics of the Last Marine Surge on Peleliu

On 27 September, tank-reinforced 2/321 and 3/321 attacked southward into the Umurbrogol Pocket from 321st Infantry Trail.

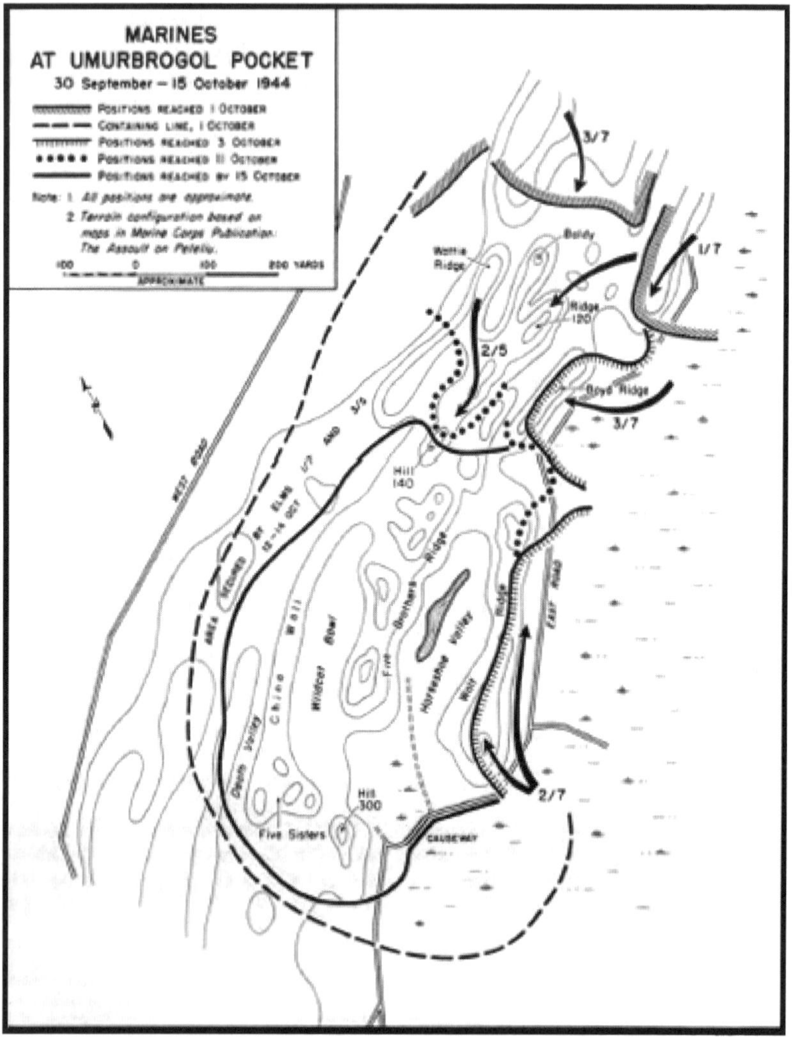

Map 13.3: The Marines' Tightening of the Pocket
(Source: "The Approach to the Philippines," by Robert Ross Smith, U.S. Army in WWII Series [Washington, D.C.: Ctr. of Mil. Hist,, 1996], chapt XXIV, map 22.)

One battalion followed the ridges overlooking the East Road, and the other Baldy and Wattie Ridges.[4] Neither enjoyed the expected degree of success. Then, General Rupertus had 3/7 resume the southward push from the Army's deepest penetration—phase line X-ray. Meanwhile, his Marine units maintained their positions along the western and southern sides of the Pocket.

> On the same day [3 October 1944] the 3rd Battalion, 7th Marines, . . . seized the eastern slope of Boyd Ridge, north of Walt Ridge. The next day, 4 October, elements of the 3rd Battalion, 7th Marines, secured a temporary foothold on the northern end of Ridge 120, but were forced to withdraw through a draw lying between Boyd Ridge.[5]
>
> — "Approach to the Philippines," by Smith
> U.S. Army Center of Military History, 1996

As the Marines drew closer to the southern part of the Pocket, enemy resistance increased

> About 1400 [7 October], six tanks with two companies of the 3rd Battalion, 5th Marines, attempted a second push. Company L moved northwest into Wildcat Bowl and Company I started into Mortimer Valley. No penetration could be made at Wildcat.[6]
>
> — "Approach to the Philippines," by Smith
> U.S. Army Center of Military History, 1996

Then, the north sides of those bowl-adjacent ridgelines had to be reattacked. (See Map 13.4.)

> On the 9th [of October], the attack was resumed on the northern side of Wattie and Baldy Ridges, where no firm foothold could be secured. Patrols of the 2nd Battalion, 5th Marines, probed into a deep gulch between Wattie and Baldy Ridges, finding there the bodies of twelve 321st Infantry soldiers, grim reminders of the previous efforts of the 3rd Battalion, 321st Infantry, in the same area during the period of 25-27 September.[7]
>
> — "Approach to the Philippines," by Smith
> U.S. Army Center of Military History, 1996

While the center of Momiji Plateau would be hard to enter, the Marines' capture of its peripheral ridgelines made its penetration later possible. (See Figure 13.1.)

On 10 October elements of the 2nd Battalion, 5th Marines, secured Wattie, Baldy, and 120 Ridges and pushed south the next day to gain the top of tactically important Hill 140, lying immediately north of the northern end of Five Brothers Ridge.[8] [Look back at Map 13.2.]

— "Approach to the Philippines," by Smith
U.S. Army Center of Military History, 1996

Figure 13.1: 2/5 Marine Coming Ashore at Peleliu

(Source: "2nd Battalion, 5th Marines Land at Peleliu," a pocket history, n.d., by 1stLt. B.B.McBreen, Project Leatheneck, Camp Pendleton, CA, artwork by CWO2 C.G.Grow)

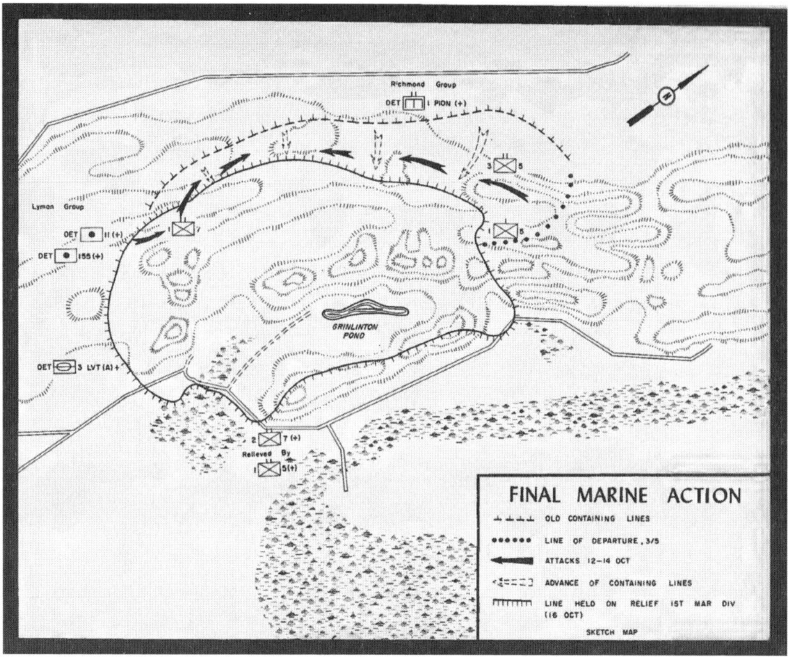

Map 13.4: Final Efforts of the Marine Units on Peleliu
(Source: "Western Pacific Operations," by Garand and Strobridge, Historical Branch, G-3 Division, HQMC, 1971, p. 247.)

As Chesty's boys had earlier in the fight, the other two Marine regiments gave it everything they had. But, in the process, the Pocket was greatly diminished. And 1st Marine Division would go on to win the pivotal battle of WWII on Okinawa.

> On 13 and 14 October, elements of the 5th and 7th Marine Regiments secured much additional ground on the western side of the central pocket between the third and X-ray phase lines, terrain so rough and broken that it was deemed useless as a route of attack into the Umurbrogol. Two days later these elements were pulled out, having cleared an area some 700 yards long north to south and as much as 200 yards east beyond the earlier containing lines above West Road.[9]
>
> — "Approach to the Philippines," by Smith
> U.S. Army Center of Military History, 1996

Hardest Fighting Was for the Horseshoe Region

The southeastern side of the Umurbrogol Pocket stretched all the way down the mountain. That may have been due to central-tunnel access through sea level caves.

> Throughout 7 October, 3/5 kept the Japanese occupied in the southeastern perimeter of the pocket. Following a heavy preparation by mortars and 105mm guns. Company I, supported by six tanks of the [Army's] 710th Tank Battalion, advanced northward and entered the Horseshoe. Both infantry and tanks raked suspected enemy positions with fire, especially those along the lower slopes of Walt Ridge on the right and Five Brothers to the left. The Marines of 1/5 on the crest of Walt Ridge gave fire support to the advance. In the course of the day, a fire team of Marines was assigned to protect each tank. The armor was also supported by two LVT flame-throwers and a platoon of the 1st Engineer Battalion. The total advance of Company I for the day amounted to 200 yards. It represented the furthest in-road into the Umurbrogol pocket from this direction. The attack had successfully reduced Japanese cave positions that had thwarted earlier advances, though their seizure was only temporary. When the tanks ran short of ammunition later in the day and had to withdraw, the infantry also had to pull back. An attempt by Company I to bypass the Horseshoe and penetrate into the valley separating the Five Brothers from the China Wall . . . drew such fierce resistance that the attack never really got off the ground.[10]
>
> — "Western Pacific Operations"
> by Garand and Strobridge
> Historical Branch, G-3 Division, HQMC, 1971

Pocket Defenders Were Far from Giving Up

A key part of the new Japanese "retreat combat" strategy then occurred outside the cordon—ground raids on Yank supporting arms.

Any idea that the Japanese contained in the Umurbrogol

> were the disorganized remnants of the island garrison was dispelled by captured orders which were interpreted to show that, as late as 1 October, the enemy still was well organized and determined to take full advantage of . . . a series of OPs [observation posts], a mobile reserve of company strength, and close-combat units specializing in night infiltration and combat. These units had been organized specifically to destroy American tanks, LVTs, mortar positions, and other important targets.[11]
>
> — "Western Pacific Operations"
> by Garand and Strobridge
> Historical Branch, G-3 Division, HQMC, 1971

Overrun tunnel entrances provided a way for Umurbrogol Pocket defenders to get behind their attackers. Then, they went after American armor and artillery. Because the battle space was now three dimensional (one below ground), it had no more front lines.

After 81st Division Took Control of the Battle

While 321st Infantry was still attached to the 1st Marine Division, the Army "direct-fired" its artillery into cave openings before rifleman approached them behind sandbags.

> [Around 19 October,] 321st Infantry, gained 25-50 yards along the ridges west of Hill 140 with the aid of 75mm pack howitzer fire and the now omnipresent sandbags. The battalion's men, prone on the ground, inched their sandbags forward with rifle butts or sticks, laboriously expanding a hold and almost realizing the infantrymen's dream of portable foxholes.[12] [Look back at Map 13.2.]
>
> — "Approach to the Philippines," by Smith
> U.S. Army Center of Military History, 1996

After the 81st Infantry Division assumed responsibility for all operations on Peleliu, a few more imaginative maneuvers took place.

> The first day of operations [around 20 October] under the control of the 81st Infantry Division was spent principally

> in reconnaissance, and on the morning of 21 October, after another napalm strike, the 1st Battalion, 321st Infantry, resumed the attack. Company A, on the right, gained almost 100 yards along its front during the morning, progressing farther toward the north end of China Wall and the northwest corner of Wildcat Bowl. . . . In the 2nd Battalion area a volunteer patrol fought its way back up to the summit of Brother No. 1, *practically by infiltration,* late in the afternoon. Company E rushed forward to consolidate this hold and a human chain was formed to pass sandbags up the hill. By dark, sandbag fortifications were completed and preparations were under way to attack south again on the next day. [Italics added.] [13]
>
> — "Approach to the Philippines," by Smith
> U.S. Army Center of Military History, 1996

But, the most cautious way of collapsing a holdout position soon became too tempting for the Army. Its predisposition toward overwhelming firepower would then dictate how to cross the Umurbrogol summit. Its heavy weaponry was taken to the north end of the plateau and shot into every available cave mouth. Concurrently, a row of infantrymen inched their way southward, further blasting and sealing every opening. Their advance resembled the gradually encroaching trenchlines of WWI.

Atop that hard coral, GIs had been unable to dig in. So, the mostly crawling soldiers pulled a few sandbags along for cover. Before long, sandbags became an accepted offensive tool.

14 Scramble for the Pocket

- What made the Umurbrogol Pocket so hard to take?
- Were the enemy positions connected in any way?

Many U.S. soldiers got hurt in that close-quarters fighting.

(Source: "Bloody Beaches: The Marines at Peleliu," by B.Gen. G.D. Gayle, Marines in WWII Commemorative Series, p. 10.)

How Much Had the U.S. Cordon Decreased in Size

The so-called "Pocket" was the Japanese occupied part of the Umurbrogol after the Marines had seized all but one of its sides. Mostly atop the massif's southern crest, this holdout area was a fairly flat valley-shaped region with northerly running ridgelines along either flank. While one ridge consisted of the China Wall and Wattie Ridge, the other included Five Brothers, Hill 140, Ridge 3, and Ridge 120 (as connected to Baldy Ridge). (Look back at Map 9.2 and then at Map 14.1.)

At the valley's bottom end was the oval-shaped Wildcat Bowl

and a high hogback ridge that ran perpendicular to the others. Comprised of Five Sisters, Hill 300, and the saddle in between, this hogback came to be known as Bloody Nose Ridge. Steep and unassailable from the airport side, its reverse slope dominated southern plateau with cannon and machinegun fire. Along both sides of the bowl were tiny interconnected bastions.

> Along the 370m of the central Omleblochel ridge and its parallel valleys, IJN records document the construction of 608 individual caves and tunnel systems, a figure that does not include foxholes, sniper positions, and emplacements.[1]
> — *Journal of Conflict Archaeology,* 2012

Within that same Pocket region, one U.S. Park Service study describes "uneven terrain . . . punctured by concealed sinkholes up to 30m deep." An official account of the battle confirms a "plateau of coral sinkholes."[2] These naturally occurring perforations in the high valley's surface made below-ground access to most strongpoints possible (as at Biak). The man-made bunker complex at the Amphitheater—the imposing anchor of Iwo Jima's MLR—may have gotten its inspiration from this Wildcat Bowl setup. (Look back at Figure 6.3.)

Overview of the Pocket Closure

While the Marines had initially established and then tightened this "last-stand" noose, its final closure by the U.S. Army involved more than just "mopping up"

> Admiral Fort . . . promptly sent a communiqué to the effect that Peleliu had been secured and that the 1st Marine Division would be withdrawn to Pavuvu, the assault phase of Operation Stalemate II complete. Over the course of D+31/32, the 321st RCT relieved the 5th Marines while the 323rd RCT relieved the 7th Marines. A number of Marine units (including the 1st Amphibian Tractor, 3rd Armoured Amphibian Tractor and 1st Medical Battalions) stayed on to support the 81st Infantry Division in a battle that lasted another six weeks. Japanese defences were now concentrated in individual positions around [Old] Baldy, Hill 140,

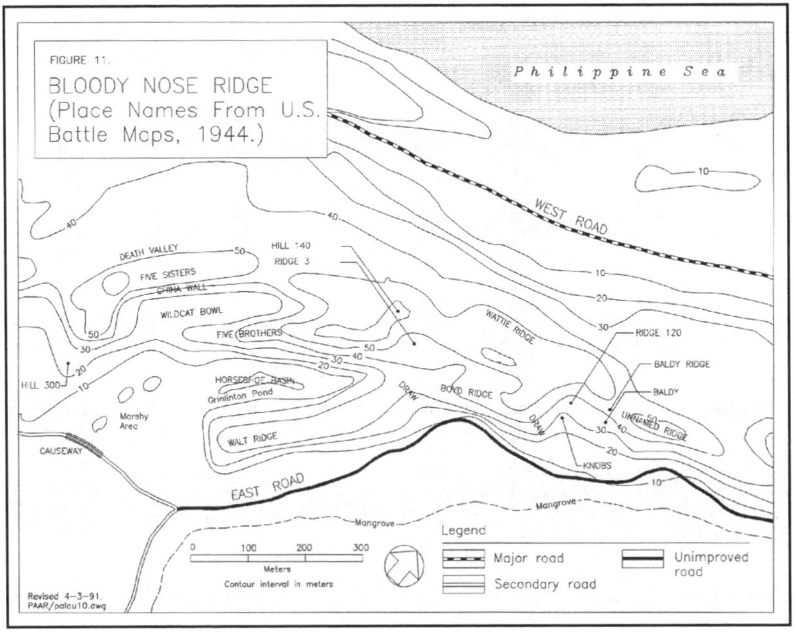

Map 14.1: Terrain atop the Momiji Plateau
(Source: "Preliminary Draft -- 1/91 (of Peleliu study)." Botany, Dept., University of Hawaii, January 1991, fig. 11.)

Five Brothers, Five Sisters and the China Wall. The Army continued to pound the Japanese, and reduce each position carefully with intense preparatory work. The 321st RCT continued the attack and took the Five Brothers and entered the Horseshoe on 23 November. 323rd RCT (under Colonel Arthur Watson) took Hill 30[0] and Five Sisters and after assuming the main responsibility for finishing the attack, started their assault on the China Wall, only yards from Nakagawa's command post in what was to be the last Japanese position on the island to fall. Engineers constructed a ramp to allow tanks and flame-throwing LVTs to fire directly onto the last Japanese defences now only a couple of hundred yards square. On D+70, Colonel Nakagawa sent one last message to Koror advising them that he had burned the 2nd Infantry Regiment's colours and split his remaining 56 men into 17 groups with orders to attack the enemy wherever they found them. That night, 25 Japanese were killed attempting to infiltrate American lines and the following morning, a prisoner confirmed that Colonel Nakagawa and Major General Murai had both committed ritual suicide in their command post. On the morning of D+73 (27 November), elements from the north and the south met face-to-face near what was Nakagawa's final command post. Colonel Watson reported to Major General Mueller that the operation was over, although mopping up would continue for some time to come.[3]

— "Battle of Peleliu," *History of War*

Who Had Done Most of the Corps' Fighting

While infantry leaders often get medals for things their subordinates do collectively, recent enlistees on Peleliu were awarded a disproportionate number of bravery citations. Of the nine U.S. Naval Service MoHs handed out, six would go to E-1's and E-2's. Of the 66 Navy Crosses bestowed, 24 would be earned by those lowest pay grades.

Considered the least knowledgeable members of most Western armies, those bottom ranks were also the most forward deployed. On Peleliu, such an assignment carried with it extra challenges. The terrain was so compartmentalized that most enemy contacts were initially made by a fireteam or buddy team. To their members should have gone the most comprehensive training.

America's infantry commanders did what they could to limit casualties on Peleliu, but a few of those efforts would later come into question.

Details of Pocket Reduction

The more firepower-oriented Army would have liked to line up all cannons and gradually move southward—filling every cave with a standoff shell, burning diesel fuel, and then handheld explosives. But, so doing did nothing to limit the plunging fire from Five Sisters and Hill 300. (See Map 14.2.) So, such a slow-moving assault was no longer ideal. Had the Marines still been on the island, they might have tried a secretive raid on Old Baldy. But, such a move would have been fully exposed and then counterattacked from inside the mountain. 323rd Infantry was eventually to capture that pesky peak in a more conventional fashion.

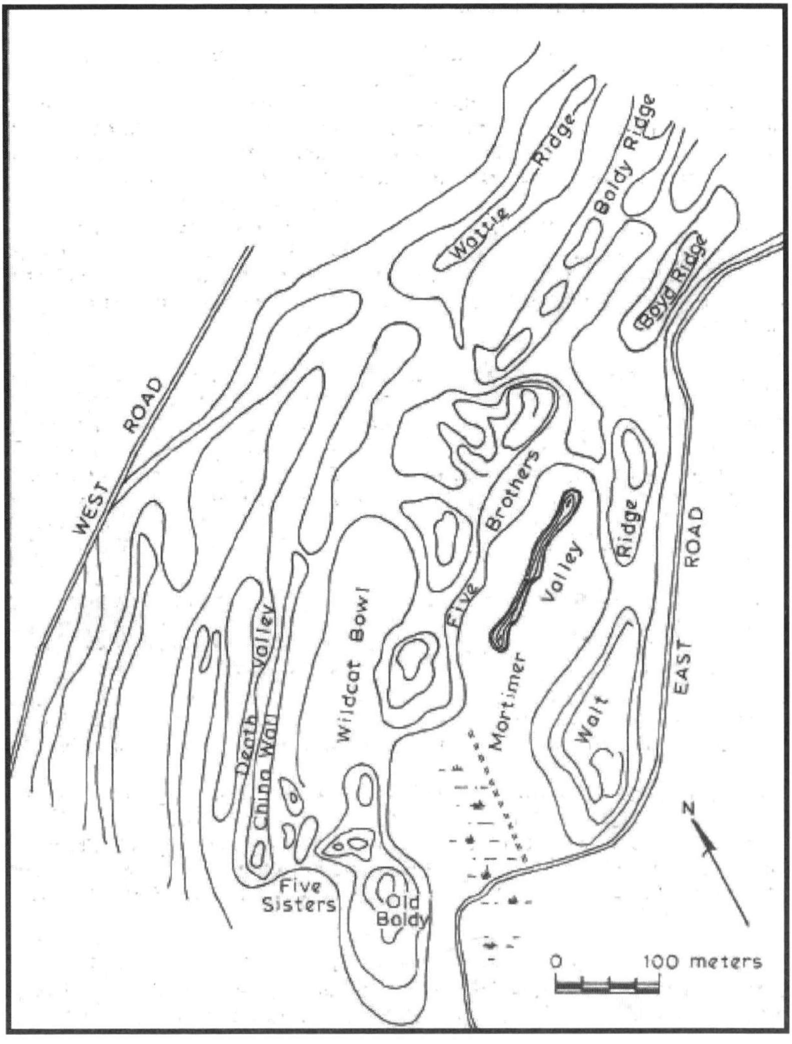

Map 14.2: Key Features in the Final Pocket
(Source: Denfeld 1981: 32, in "WWII Battlefield Survey of Peleliu Island...," by Knecht et al (Washington, D.C.: Nat. Park Service, 2012.)

> The Eighty-first began a methodical tightening of its lines around the pocket on October 21. Not without difficulty it captured the northern hills called the Five Brothers, one by one. In Horseshoe Valley, west of Walt Ridge, it succeeded in cutting the Japanese off from their water supply at Grinlinton Pond. (This accomplishment was offset by heavy rains that replenished the Japanese water supply.) On November 2, the 323rd RCT captured the Five Sisters, including Bloody Nose Ridge [the saddle between them and Hill 300] against which the 1st Marines had hurled itself so long ago. Japanese fighting strength was now reduced to 350 men. Little action occurred on either side between November 4 and 9 as a typhoon swept over Peleliu. On November 13, the assault and occupation of Peleliu was again announced, this time by General Geiger.
>
> That same day, the two [U.S. Army] infantry regiments began the final drive. On November 24 . . . a last [Japanese] radio message was sent to Koror advising that the 56 remaining combatants had organized into seventeen raiding groups and would hide in the jungle and caves to harass the enemy. . . . It had taken U.S. forces almost 2 1/2 months to complete what began as a four-day operation. The Eight-first Division did not declare an end to organized resistance until November 27.[4]
>
> — National Park Service Study I

Implicit in that enemy stratagem was a way to move unseen from the top of the Umurbrogol to the mangrove swamps below. With thousands of U.S. troops now swarming over much of the barren mountainside, it couldn't have been through short-range infiltration. Tunnels must have been involved. The Japanese were still working on their alternate mission—creating an unacceptable number of American casualties while delaying a homeland invasion. To finally remove such lethal opposition, the U.S. Army also relied on its lowest-ranking personnel.

> Nakagawa would make the [U.S.] soldiers pay for every inch of Peleliu. It was an infantryman's fight. It was the individual rifleman, armed with satchel charges, grenades, and explosive devices, who finally reduced the Umurbrogol to a pocket 900 yards long by 400 yards wide. . . .

> The 321st and 323rd RCTs moved in to continue reducing the Umurbrogol Pocket. The 306th Engineer Battalion was able to erect a coral ramp from the bottom of Wildcat Bowl to the top of China Wall so artillery and supplies could get to the soldiers fighting there.
>
> The 81st Division inched its way forward, eliminating cave after cave.[5]
>
> — *Warfare History Network,* 4 January 2019

Unfortunately, the most logical way of mopping-up soon became productive to the point of moral excess. Instead of garden pests, U.S. forces had been subduing hostage-holding foreign soldiers. But, church-going ambush victims can unintentionally get the two confused.

> As a means of reducing the last Japanese defenses on Peleliu, ingenious Soldiers set up fuel tanks in covered positions about 300 yards from the Japanese caves, then hooked up a hose to the tanks and poured oil into the most prominent enemy caves. This oil was ignited by white phosphorus hand grenades lobbed into the caves after the spraying. This method yielded good results and henceforth became an effective improvisation.[6]
>
> — National Park Service Study II

Nature of Enemy Resistance Deep Inside the Pocket

Front lines were of little use to U.S. units trying to clean out the Pocket. Japanese fighters started showing up behind the cordon, in supposedly secure areas.

> The Japanese, even before their counterattack against Five Brothers, had sent infiltraters *[sic]* into action on the southwest side of the pocket, chiefly by means of tunnels and connecting caves, to points some 300 yards outside the recently established containing lines of the 1st Battalion, 323rd Infantry (1st Bn Rpt, 20 Oct-3 Nov 44).[7]
>
> — "Approach to the Philippines," by Smith
> U.S. Army Center of Military History, 1996

How a Throwback to WWI Then Helped the Army

The 81st Division's degree of caution during the Umurbrogol Pocket assault soon took on aspects of a former world war. The hardness of the ground and lack of cover had greatly influenced the decision. (See Figure 14.1.)

> It remained for the 81st Division, following relief of the Marines, to develop a technique with such ingenious refinements [beach sand and ammo transported to the top of the mountain by overhead cable system] as to make the sandbag into something closely resembling an offensive weapon, in which capacity it played a crucial part in the final reduction of the Pocket (Hough, "The Assault on Peleliu," pp. 164-65).[8]
>
> — "Approach to the Philippines," by Smith
> U.S. Army Center of Military History, 1996

Figure 14.1: The Fighting Was More Intense Than on New Guinea
(Source: "Red Arrow at War," by Michael Gnatek, poss. ©, from this url: https://www.nationalguard.mil/Resources/Image-Gallery/Historical-Paintings/Heritage-Series.)

After each cave was shelled, a row of defensive positions was gradually shifted forward.

> On 23 October, the only significant advance was made by Company E, 321st Infantry, which proceeded south along Five Brothers Ridge to secure Brother No. 4 and fortify it with sandbag emplacements. . . . On the 24th, the 1st Battalion gained 50-75 yards, laboriously inching forward its sandbags; the 3rd Battalion continued sealing caves along both sides of Mortimer Valley.[9]
>
> — "Approach to the Philippines," by Smith
> U.S. Army Center of Military History, 1996

> More active operations were resumed on 2 November when the 2nd Battalion, 323rd Infantry, attacked Five Sisters from the southwest; elements of the 1st Battalion moved north along the ridges west of Death Valley, pursuing the last Japanese from South Pocket; and part of the 3rd Battalion started south along ridges and broken terrain at the north end of China Wall. The latter's attack gained only 25-50 yards as Company L sandbagged its way forward literally inches at a time. . . .
>
> Attacks at the Umurbrogol started in earnest again on 13 November, with the 1st Battalion, 323rd Infantry, pushing eastward from the ridges west of Death Valley and the 2nd Battalion moving north into that gorge and Wildcat Bowl. The 1st Battalion's attack gained little ground, and that only by sandbagging almost every foot of the terrain.[10]
>
> — "Approach to the Philippines," by Smith
> U.S. Army Center of Military History, 1996

To provide enough pre-assault bombardment, roads were built atop the Umurbrogol for the Army's towed artillery—to include one up the China Wall itself.[11] The southern end of the Pocket was blanketed with napalm by aircraft from Peleliu's own airfield. And then eventually, the 323rd Infantry was able to seal up the final set of openings in the ground. While highly deliberate and well planned, the U.S. Army's sweep of the enemy's last-stand position still proved costly. That's because the Pocket had so many subsurface passageways through which the quarry could ambush his tormentor.

Precursor to Future Fights

Peleliu may have been the first organized example of defenders resupplying themselves with what they could steal from a better-endowed attacker. On Iwo Jima and Okinawa, such activity would become a regular part of "retreat combat."

> Night-time infiltration raids may have had the secondary result of killing American soldiers [on Peleliu], but it is clear that towards the end of the battle the primary purpose was to steal their water and food.[12]
>
> — *Journal of Conflict Archaeology,* 2012

And, to this day, American expeditionary forces flaunt their wherewithal by practicing a continually higher-tech version of 2GW.

Part Three

What Was Learned That's Still Useful

"I'll be Home for Christmas, If Only in My Dreams"
— Bing Crosby song from 1943

(Source: Attributed to Bing Crosby.)

15 Peleliu Lessons

- Had a more complete reconnaissance been possible?
- What constitutes the biggest threat to a on-line unit?

Peleliu was eventually fire-bombed from its own airfield.

(Source: "Bloody Beaches: The Marines at Peleliu," by B.Gen. G.D. Gayle, Marines in WWII Commemorative Series, from public domain cover art by Tom Lea.)

The Landing Had been Difficult

Four hospital ships had lain offshore at Peleliu. They would receive 2,343 wounded Americans in the early days of the invasion. The average life span for Marines debarking the first-wave of landing craft was 46 seconds.[1]

The landing force commander, Maj.Gen. Rupertus, had hoped the island would fall in four days. It ended up taking 73 days of vicious, often hand-to-hand, fighting. The overall U.S. casualty rate—80% in some units—was the highest of any amphibious assault in the Pacific.[2]

There Had Been a Few Logistics Problems

At Peleliu, 1st Marine Division mounted the improvised Mark 1 flame thrower system on the thin-skinned LVT-4. Vehicle vulnerabilities were then to limit system effectiveness.[3]

The seizure of separate bunkers and cave openings typified the struggle for Peleliu. Most U.S. advances were slow, siege-like operations in which every emplacement had to be isolated from those around it and then individually reduced by close combat. Though based on the "blow torch (flame) and corkscrew" (satchel charge)" method, they soon depended more upon tiny-element maneuvers through the microterrain.[4] And grenades became the most useful ordnance.

> U.S. records show an expenditure on Peleliu of 118,000 grenades, 150,000 mortar bombs, 123,000 artillery shells, and a staggering 15.5 million rounds of 0.30, 0.45, and 0.50 calibre bullets by the Marines and [Army] Infantry alone. To this may be added the 5849 tons of ordnance fired by the Navy and the 806 tons of bombs dropped from aircraft.[5]
>
> — *Journal of Conflict Archaeology,* 2012

> "Postwar statisticians calculated that it took U.S. forces over 1500 rounds of ammunition to kill each Japanese defender and that during the battle, the Americans expanded 13.32 million rounds of .30-calibre, 1.52 million rounds of .45-calibre, 693,657 rounds of .50-calibre bullets, 118,262 hand grenades, and approximately 150,000 mortar shells."[6]
>
> — Peleliu Dive Adventures website

> Navy ships anchored offshore fired almost 6,000 tons of shells. Navy and marine aircraft dropped at least 800 tons of bombs. In their month on the island, the Marines expended almost 16 million rounds of ammunition, including 116,000 hand grenades.[7]
>
> — *Archeology Magazine,* July/August 2019

Moral Issues Had Come into Play

The Japanese had made the taking of Peleliu as costly in human life as possible. How such a lethal adversary was subsequently handled would be fully understandable under normal circumstances. The landing force had been badly hurt, and the inland defenders didn't like to surrender.

But, there had been thousands of forced laborers also present on Peleliu. Making too little provision for their welfare would today violate certain amendments to the Geneva Conventions. The final clearing of the Umurbrogol Pocket in 1944 more closely resembled an extermination exercise than mopping up. All openings in the ground were filled with fuel, set on fire, sealed with explosives, and then likely bulldozed.[8]

> On average 1,589 rounds of heavy and light ammunition was fired for each Japanese death. No one knows how many Japanese soldiers were sealed in the caves. . . . Many of the prisoners were Korean laborers.[9]
>
> — "Battle of Peleliu," *Facts and Details*

U.S. Capabilities from a Worldwide Perspective

By too heavily depending on established procedure, Western military bureaucracies unknowingly perpetuate the *status quo*—resist any change, however useful. Their budgetary needs are more easily met by appearing to do well at all things. So, to put every action into the most favorable light, the U.S. service branches maintain small armies of public affairs and legal assistance personnel. The wartime potential of bottom infantry echelons are the least visible. Tiny element inadequacies are seldom shared with high-level commanders to safeguard the reputation of intermediate leaders and the organization as a whole. A chronic shortfall in infantry squad, fireteam, buddy team, and rifleman tactical technique results.

Cultural Differences on Earth

Many Americans seem unaware that the predominate culture in the Eastern World is—in many ways—opposite to their own. Frequently described as "bottom-up," that culture solves many of its problems through the collective effort of ordinary citizens. For example in India, there used to be few street lights or stop signs. At any intersection, Indian drivers simply cooperated enough for everyone to get through it safely. This *"gung ho"* tradition predates Communism by thousands of years, and may have its origins in the more massive populations of Asia. Meanwhile, Western World solutions to the same problems have come from elected or appointed officials.

While all Asian Communist armies serve an authoritarian regime, they still allow their infantry echelons to follow this bottom-up methodology. What works tactically for small units drives large-unit operations, and successful operations lead to overall strategies. Entry-level Oriental infantrymen are also provided with more movement options than their American counterparts. That's why frontline Asian fighters tend to display more battlefield initiative.

Such realities of an Eastern war makes "attention to detail" an absolute necessity for all Western armies. Tiny events along the front lines of any battle may be the best indicator of enemy intentions. And what the U.S. rifleman sees and reports may make the difference between winning or losing the engagement. During the Korea War era, U.S. Marine NCOs used to play an interesting game around the campfire at night. One man would request some military detail, and whoever knew it would receive all the quarters in the pot. While most questions focused on weapon idiosyncrasies, some had to do with the tactical insights from WWII. For example, Marine leaders in Korea crawled while checking their lines at night to keep the precise location of each hole secret. By the time the Vietnam War rolled around, those low-profile visits had been replaced by telltale silhouettes.

How a Tactical Shortfall Could Have Occurred in America

The well-respected principle of Occam's Razor is this: "Of two explanations that account for all the facts, the simpler one is more likely to be correct." That principle applies to a wide range of disciplines, including religion, physics, and medicine."[10] It may also apply to military matters. No tactical shortfall could have survived the Peleliu clinic without some other issue blocking its rectification. That other issue may be trying to make money and win wars at the same time. The modern-day Pentagon works almost entirely on military contracts, and the old-time War Department may have done the same thing.

The Christian version of Occam's Razor is less tactful: "The lie gets started where something inherently simple is made more complicated." But, armies never lie, they just put every occurrence into the most favorable light. A good modern-day example of Occam's Razor within the U.S. Military may be how to decisively win the next war. A conclusive victory would require initially controlling all enemy territory. In a large war zone, that's only accomplished through U.S. infantry squad outposts. No amount of aerial surveillance or punitive bombardment will get the job done. Yet, America's professional military journals spend little, if any, time talking about how to create "truly light" (or self-sufficient) infantry squads. Doing so in-house would cost nothing, but neither would it stimulate the economy.

Throughout history, U.S. veterans have advised returning to the "basics." After the Battle for Guadalcanal, the 1st Division Commander, Maj.Gen, Vandegrift, went so far as to say that "jungle warfare against the Japanese is a question of going back to the tactics of the French and Indian days."[11] That's why all possible subcategories of "shoot," "move," and "communicate" must be considered while assessing what was learned from the Battle of Peleliu.

Western top-down cultures tend to focus on "the big picture," whereas their Asian counterparts study its composite parts. Any lack of foresight by U.S. planners on Peleliu can likely be traced to this societal difference. Too little "attention to detail" had come into play on important issues.

What U.S. War Planners Had Perceived on Peleliu

From high overhead, central Peleliu had looked like a series of well-vegetated, rolling hills. "[T]ropical forest subdues its surface relief."[12] Not until the central hill-mass was defoliated through U.S. bombardment did its precipitous and broken nature become apparent. Then, the Umurbrogol became a nearly insurmountable obstacle.

To fight successfully in the Eastern World, the unique nature of the terrain must be ascertained. The U.S. Army had failed to do so at Biak, and U.S. Naval components followed suit at Peleliu.

Peleliu's Oversights Will Help GIs to Win the Next War

Unlike U.S. proponents of the steamroller strategy, Asia's military leaders carefully study their adversaries. They have known for years that American field commanders depend upon maps and aerial photographs that reveal little, if any, of the minor ripples in the ground. For a rifleman in heavy combat, a one-foot rise can mean the difference between life and death. So too can a spider hole occupant demobilize a tank without its infantry guards ever knowing what happened.

Peleliu's most glaring big-picture disconnect was the difference between its apparent and actual surface composition. Then, through the Western version of momentum, the landing force had incurred excess casualties. Within the Eastern World, momentum is generated a different way—no longer through speed or irresistible force, but through gradually larger encounters. The reason is inescapable. Through a bottom-up culture, Asians pay more attention to microterrain. Until high-level American leaders realize its importance, their war machine will continue to underperform. It will go where it wants as in South Vietnam, but the land it crosses will remain in enemy hands.

The Peleliu experience provides a useful overview of what American troops will face during any future conflict in the Pacific. As Winston Churchill said, "Those that fail to learn from history are doomed to repeat it."

Too Few Ground Irregularities Will Be Foreseen

Aerial or satellite photographs have great difficulty distinguishing the tiny features (ditches, mounds, etc.) that blanket rural landscape. Yet, to dismounted infantry, those tiny ground irregularities provide the only way to operate safely.

So, before the Peleliu invasion, U.S. planners worried too little about the "danger area" (naturally occurring kill zone for indirect-fire shrapnel and grazing machinegun bullets) that covered the entire south end of the island.[13] Most of the airport clearing had been so hard and flat as to offer no cover at all. Nor did those planners notice the tiny rise beyond the beach that was to so impede the left side of the invasion.

> [T]he leading elements had not advanced a hundred yards inland before they found themselves confronted by a most formidable natural obstacle which showed on none of the advance maps or intelligence reports: a long, rugged coral ridge about 30 feet in elevation, its precipitous face honeycombed with caves and dug-in positions. All initial assaults were repulsed, even after the tanks had been brought up. Troops attempting to storm the northern portion stumbled into a wide, deep anti-tank trench, dominated by the ridge itself and cleverly enfiladed, where many of them remained pinned down for hours.[14]
>
> — "Western Pacific Operations"
> by Garand and Strobridge
> Historical Branch, G-3 Division, HQMC, 1971

U.S. leaders had considered Peleliu too heavily occupied for a full reconnaissance. But, discovering its surface limitations would have required commandos with a straight-leg infantry background. The lesson is clear. All future objectives must first be inspected —from within—by people who know what to look for.

> With this somewhat speculative intelligence, the planners had to be content, since it would be manifestly impossible to send pre-invasion patrols ashore [at Peleliu] on a land mass so small and so strongly held.[15]
>
> — *The Assault on Peleliu,* by Hough
> History Division, HQMC, 1950

The NVA (North Vietnamese Army) and VC in Vietnam routinely reconnoitered the interior of every American base they attacked.[16] Those U.S. bases were just as well populated as Peleliu. But *ninjutsu*-trained infiltrators have little trouble with such a mission.

Some Terrain Will Be Defended from Afar

On Peleliu, large guns on the forward face of the Umurbrogol could take every corner of the airport clearing under precision bombardment. Some of the landing beach shelling was also of the plunging variety. Yet, most was from no fewer than six sand spits that stuck out into the ocean. While such shoreline anomalies can be caused by a prevailing wind, those on Peleliu were so evenly spaced as to suggest a little bulldozer work. Either way, invasion planners had failed to realize their combined threat. The American-drawn schematic on Map 4.4 shows flanking fire from only two spits, whereas the original Japanese version—Map 4.3—depicts final protective fires from five oceanside strongpoints.

> [F]or whatever reasons, enemy installations enfilading both flanks of the landing beaches were still operative when the assault waves went in and inflicted casualties to personnel and materiel so severe as to hamper seriously the first day's operations.[17]
>
> — *The Assault on Peleliu,* by Hough
> History Division, HQMC, 1950

The prevailing winds on Peleliu Island (also known as Bliliou) were from the wrong direction to create spits like these. Its land mass ran from southwest to northeast, and a sand spit is created when the prevailing wind hits the coast at an oblique angle.[18] On Peleliu, such a breeze would have had to come from due west. It didn't.

> Palau [the Pacific nation] is subject to two wind patterns during the year: the northeast trades from December through March and the southwest monsoons . . . from June through October.[19]
>
> — Hawaii University study, 1991

Nor had the precise height and composition of those shoreline projections on Western Peleliu been determined. The northernmost "spit of land that jutted out" ended up being the 30-foot-high headland to a rock promontory. This location would later come to be known as "The Point." Holes had been blasted into its coral outcropping to accommodate a 47mm (1.85-inch) gun and six 20mm cannons. These enemy gun emplacements were then fully encased in coralline limestone, leaving only small slits through which to fire sideways at the debarkation zone. According to one source, "similar positions" had been crafted all along the two-mile stretch of landing beach.[20]

Enemy Fire Will Come from Unexpected Directions

At Tarawa atoll in the Gilberts, Japanese machineguns on a rusting hulk in the harbor had done much of the initial damage. Invasion participants were too busy dodging fire from the beach, to notice any from the rear.

During the Peleliu landing, the most devastating antitank fire had come from the Point. There had also been well-aimed shots from an islet at the south end of the beach. But, those were not the Marines' only problem. With a range of 2000 meters, heavy machineguns at the airport's service area could have reached either location. An inland sweep would later be hurt by long-range machinegun fire.

Foremost among Peleliu's many warnings is that "any linear formation requires fully secured flanks." The cost of ignoring this rule will continue to be too many casualties.

U.S. Fire Superiority May Be Impossible to Establish

Of the non-tactically arranged Navy caves at the northern end of Peleliu, the 5th Marines Regimental Narrative reported the following: "Tank guns, firing point-blank directly into the caves and tunnels, did not even temporarily cause the enemy therein to cease fire."[21] In other words, fire superiority had been impossible to establish before the ground assault.

Enemy Bunkers Will Be Fully Fortified

At Biak, the enemy had stopped building coconut log emplacements. During the Peleliu struggle, some were in caves with cement gun ports and steel covers. Others were in camouflaged pillboxes of reinforced concrete. And all were connected by tunnel, instead of trenchline.

There Will Be a Hostage Element

There were thousands of kidnapped Koreans and Okinawans on Peleliu when the Americans landed there in 1944.[22] And at

Biak, there had been Javanese, British Indian, and Formosan slave labor.[23] After any future enemy acquisition, there will minimally be impressed inhabitants. All enemy-controlled locals and foreign workers will constitute hostages who rate protection under the Geneva Conventions.

Withdrawing Defenders May Join a Reserve Force

Reserves had existed at the center of Umubrogol Mountain. (Look back at Map 4.4.) Through the many tunnels, those Reserves could then run spoiling attacks and then defend the last-stand location.

Some Counterattacks Will Come from Unlikely Locations

Map 7.4 reveals places at which four major counterattacks originated: (1) southern end of Umurbrogol; (2) Horseshoe area; (3) beside an airport runway; and (4) near the southern lagoon. Those attacks most likely came from an underground passageway. A non-Umurbrogol-connected tunnel could have existed at the south end of the island, with its excavation debris hidden along a nearby road. More than one Japanese Naval Construction Battalion had been on the island.[24]

A close-up of Map 4.2 reveals a shoreline cave near the start of the southernmost counterattack. If its participants hadn't emerged from a tunnel, they'd been hiding in this tiny lagoon-front cave.

> This passage describes the tactic of "passive infiltration" (i.e., hiding out in overrun positions in order to be in the rear of U. S. troops) which played such a prominent part in the operation.[25]
>
> — *The Assault on Peleliu,* by Hough
> History Division, HQMC, 1950

The Most Important Fights Will Be By Lone Fireteams

To avoid Western firepower, Eastern forces often "make their stand" in severely broken terrain. All the natural compartmentalization then gives their more tactically proficient small units the edge.

> . . . [L]ater the high ground itself proved a maze of fire-swept razorback ridges and barren declivities devoid of natural cover.
>
> Here, if ever, was essentially an infantry battle, with all supporting arms and troops concerted to help the man on foot in their several ways and with varying degrees of effectiveness, again mainly as dictated by terrain conditions.
>
> During the initial phases, their parts were the conventional ones they had been trained to perform. But once the fighting entered the ridges, more particularly after resistance was localized within a constricted pocket, the battle became one of improvisation: traditional Yankee ingenuity and the will to live pitted against Japanese preparation and the fanatic will to die; with each offensive thrust, often each individual cave, presenting problems peculiar to itself.
>
> It is this feature, perhaps more than any other, which characterizes the Peleliu operation in the minds of those who participated or who have studied it.[26]
>
> – *The Assault on Peleliu,* by Hough
> History Division, HQMC, 1950

Outguards Can Mask an MLR's Whereabouts

Machineguns provide the final protective lines (FPLs) that insulate any defense line from human-wave assault. That's why automatic-weapons fire from elsewhere can feign an MLR. The strongpoint belts on Bairoko, Biak, and Peleliu had been deeply outposted. While a stationary array would have warned of impending attack, roving sentries were safer. So, enemy snipers withdrew to the first row of bunkers after shooting a U.S. leader, adjusting supporting arms, or covering each other's movements by fire. The WWII Germans deployed roving sentries to better protect against infiltration.[27]

At Bairoko and Peleliu, some Nipponese snipers had *Nambus.* Even from a spider trap, they could attrit an attack force while keeping the MLR's location secret.

Peleliu Had Provided Hints of How to Cross a Tough MLR

From 19 September 1944 onward, 2/7 had likely struggled to get beyond the parallel defiles that ringed the southwestern corner of the Umurbrogol. Together, those defiles had comprised an imposing defense line. (Look back at Map 7.3.) How that most challenging of possible MLRs was handled still constitutes a step forward in the evolution of small-unit tactics. Clearly, the bazooka played a role.

> 19 September 1944. The only bazooka operator left in his company during an attack against enemy positions deeply entrenched in the coral hills north of the airfield on Peleliu, Private First Class Lilja [of 2/7] fought the fanatic Japanese with determined fury, pushing tenaciously over the rocky ground and tangled ravines to deliver his shattering fire against numerous hostile caves and emplacements which, safe from our machine guns and rifles [look back at Figure 7.2], poured a merciless rain of devastating crossfire on our troops. Fearlessly risking his life, he ranged the entire front area, carrying his unwieldy weapon and heavy ammunition load up and down the ridge despite intense heat and unceasing enemy barrages and, boldly employing his daring assault tactics in support of each platoon in turn, consistently blasted the heavily defended Japanese caves, neutralizing many strong points and materially furthering the advance of our forces.[28]
>
> — Navy Cross citation for PFC George Lilja

Within how that bazooka was deployed lay the key to defense-line crossing procedure. It was fired to one side by a lone forward-ranging operator. Just as the NVA had often put an RPG (Rocket-Propelled Grenade) man on point in Vietnam,[29] so too had the Marines stationed a bazooka man at the front of their assault formation. In this way, the enemy emplacements were given no forewarning of the decisive explosion. Since that time, all successors to the WWII bazooka have been relegated to the parent unit's rear echelon to keep as many friendlies as possible out of its backblast area. Once again, too much peacetime emphasis on safety has adversely affected America's wartime proficiency.

> 3 October 1944. Pinned down by withering hostile small-arms fire emanating from a firmly-entrenched cave while advancing single-file with a forward assault platoon against a steep, jagged cliff, Private First Class Carlton [of 2/7] refused to seek cover and, . . . succeeded in neutralizing the position with expertly placed bazooka fire, thereby enabling the advance to continue.[30]
>
> — Navy Cross citation for PFC Ted. J. Carlton

Too Much "Top-Down" Focus Has Been the Problem

After an incoming bazooka round, any cliff-side gun site would have received a visit from the most nimble Americans—normally recent enlistees. The organization's least knowledgeable members had suddenly become its most vital. Perhaps they should have been better prepared to operate independently. Would more electronic gadgetry have helped them to defeat world-class light infantrymen?

16 What Made the Taking of Peleliu Important

- Which enduring service to America had Peleliu provided?
- How did 1st Marine Division later contribute to WWII?

Within fireteam tactics lies the key to fewer casualties.

(Source: "Planting of 20,000 Flags on Memorial Day," by Ellen White, U.S. Army Center of Mil. Hist., image designator "20KFlags.jpg," possible ©.)

Issues of Interest to Modern-Day Warfighters

On Peleliu, the Japanese had partially outfoxed the invasion planners. Then, the American landing force was able to launch some viable—though still costly—countermeasures. Until the Pentagon puts more emphasis on situational detail, enemy tactics, and U.S. small-unit maneuvers, further examples of "belated success" are likely.

But, the reasons behind such a mixed outcome are still of help to present-day GIs. Peleliu was not a very big island. Yet, to the men of the 1st Marine Regiment, the 900 yards between coastline

Point and southwestern base of the Umurbrogol seemed like 900 miles. That's because they had been forced to crawl most of the way. In short, the actual battle was fought at much smaller scale than expected. While the invasion planners had focused on "the big picture," their opposition had built a defense around tiny details (like what lay hidden below ground). Senior U.S. leaders expected to take Peleliu through the combined firepower of fully supported infantry battalions. Col. Chesty Puller first discovered the truth. This "garden spot" would have to be captured one bunker or tiny crevice at time by miniscule elements. And within their separate actions would lie more sophisticated stratagems for future combat.

To win another extended conflict within Asia, Americans will again have to fight at this smaller scale—one in which microterrain maneuver and hand-held weaponry are the deciding factors. As at Peleliu, their supporting arms—however powerful—will make little difference. Most ground momentum will come from consecutive small-unit victories.

2GW will always be a part of warfare. In some cases, pre-assault bombardment may save a few lives. But, so too will more 3GW-like maneuvers. That's because how long the battle lasts will depend on the durability of the foe's command and logistics networks. Any American attempts at 3GW will at first mimic what the opposition has been doing—ambush and infiltration. But, not until all U.S. riflemen are trained for both types of war will their commanders risk a full 3GW regimen. Nor will those riflemen enjoy optimal survivability. In the absence of orders, only a 3GW-trained warrior knows what to do next.

Current Misperceptions

At present, the American public thinks it has the best trained junior infantrymen in the world. Yet, no more straight-leg U.S. ground pounders exist (they all ride). So, the microterrain is no longer any GI's domain. And, Yank special operators are so seldom required to breach a deliberate defense that they lack advanced assault technique. In other words, no part of the Department of Defense (DoD) has been keeping up with the worldwide evolution of squad tactics. Against the low-tech armies of East Asia, that's a considerable handicap.

In truth, U.S. riflemen are only the best trained in shooting their weapons and doing as told. In most other aspects of combat, they've been given only enough skill to accomplish their CO's overall scheme of maneuver. Too much low-level initiative would be counterproductive.

So, little of the current GI's preparation has to do with personally evading enemy fire. Those "snuffies" from Poughkeepsie and Stocton are either up or down depending on which large-unit maneuver they are participating in. Such an arrangement may have been good enough before the invention of the machinegun, but it has since proven costly. If that recent enlistee can't somehow avoid the streams of enemy bullets while changing location, he gets dead. And if his parent unit doesn't suffer excess casualties under similar circumstances, it fails at its mission. This way of operating may have gained the eventual victory on isolated Pacific islands, but it has less utility where there's no blockade. That's why—for America—only ties and losses have been recorded from all extended ground conflicts since WWII.

Peleliu's Ongoing Importance

Until Peleliu's lessons become part of the "corporate knowledge" of both U.S. infantry organizations, the intensity of that 1944 struggle will continue to provide insight. But, not all of those lessons can be taken at face value. It was at Peleliu that both service branches encountered their most challenging terrain in history. The beach fortifications were below ground and facing sideways on sand spits. That's why no U.S. ship or plane noticed their presence. One defense line had consisted of parallel crevasses with fortified sides. It was thus protected from both standoff fire and rapid overrun. And, the ground within the Umurbrogol Pocket had been so unobstructed and fire swept as to prompt U.S. Army units to use sandbags on offense. So doing made perfect sense at the time. From a 2GW standpoint (the "meatgrinder" approach to war), sandbags would help assault troops to cross a hard and flat surface. That's just as logical as gradually encroaching trenchlines on softer soil. But, such a deliberate a way of moving forward also increases the battle's duration, and with it the casualty count. So, alternatives must be found.

There had been another way to solve some of Peleliu's terrain challenges, but not until recently has any U.S. infantry outfit acknowledged its legitimacy. Called "subterranean warfare," U.S. Army troops now practice how to fight below ground in Korea.[1] Had the Japanese headquarters, ammunition-filled mine shafts, or Reserves cavern been reached via sub-surface passageway early on, the battle would have ended sooner. After all, Japanese Map 4.3 does show where the headquarters was, and even the tanks. But all of these strategically vital targets were so deep beneath the mountain as to be immune from standoff bombardment. The biggest modern-day "block-bluster" couldn't have touched them. Their destruction required the WWII equivalent of a satchel-charge-carrying tunnel rat from the Vietnam era. But, in 1944, sending a few people below ground was considered more dangerous than a whole unit moving upright through machinegun fire.

Not until the Pentagon comes to grips with such inconsistencies, will memories of Peleliu cease to play an important role in future U.S. combat.

GIs Will Continue to Face a More Tactically Adept Foe

U.S. chronicles are full of how innovative the Japanese became trying to win WWII. From Biak on, they no longer repelled invaders through shoreline stands or *banzai* charges. Instead, they adopted the core principles of 3GW—ambush and infiltration. They gave up certain 2GW procedures and got more "bang for their buck" out of others. But, because their intent was to force America into an armistice through unacceptable attrition, U.S. warfare experts never picked up on the shift in style. And Washington's War Department doggedly stuck with its own 2GW regimen, most notably overwhelming bombardment.

But, desperation was also to visit U.S. lines. From Peleliu onward, American forces endured continual ambush and infiltration. Something called "retreat combat" emerged as Japan's overall strategy. It enabled the user to get more productivity out of his dwindling supply of manpower and ordnance.

Eastern Defenders Will Make Full Use of the Terrain

Peleliu may have provided the first example of something most

Marine leaders have trouble imagining—nearly non-assailable real estate. Chesty Puller's "fire eaters" hadn't lost heart trying to reach the top of the southern Umurbrogol, as a few journalists have unkindly speculated.[2] They simply ran into lateral cracks in the ground that were so well fortified and closely arrayed as to be impossible to cross in a hurry. And in an attempt to comply with the O-3 phase line,[3] they had taken too many casualties to maintain T/O (Table of Organization) strength. Had there been more time and replacements available, they would have eventually penetrated that nightmarish defense line. But, with 20/20 hindsight, it would have taken nighttime infiltrators tossing grenades into every rear-side opening of each thin ridgeline. And seasoned commandos would have trouble doing that.

The most obvious lesson from Peleliu is this. For some objectives, the attacker must go around. But Chesty hadn't enjoyed that luxury. He'd been given a time-sensitive mission and a specific lane in which to operate. Any delays or 3GW-like maneuvering would have been the prerogative of superiors.

A Good MLR Not Breachable Through Conventional Combat

An ever deeper lesson may reside in this 75-year-old island campaign. It has to do with any military force that cannot or will not move beyond 2GW-like methods. No matter how well supported or mechanized an attack may be, it won't be able to penetrate a properly prepared MLR with brute force alone. That's why a U.S., NATO (North Atlantic Treaty Organization), and U.N. (United Nations) juggernaut could not enter North Korea in 1953. And why the Israelis' fully supported raid on Southern Lebanon in 2006 failed.[4]

At Iwo, Marine commanders would need "machinegun-enveloping scouts" to break through the Japanese defense lines.[5] Out of abject necessity, volunteer riflemen accomplished the infiltration aspect of 3GW. Without some lane preparation, even an aircraft-backed mechanized assault won't get far into a well-designed MLR.

The most logical choice for such a mission are the frontline infantrymen. Far fewer entry-level "volunteers" were cited for bravery on Peleliu than on Iwo Jima. Nor were there as many former Maoist Raiders in charge in the 1st Division on Peleliu and in the

3rd, 4th, and 5th Divisions on Iwo. So, not as much impromptu infiltration or 3GW-like "recon pull" occurred during the Palaus episode.

However, five out of the eight medals of honor on Peleliu would go to E-2's. So, authority had been properly delegated within its Marine outfits. Size of the assigned unit had matched the available space—one buddy team per cleft in the rock. Such a procedure saved lives. And Gen. Rupertus had been responsible for that as well.

More Small-Unit U.S. Proficiency Will Be Necessary

Chesty's boys had been given an impossible mission on the Umurbrogol, and then come very close to accomplishing it. But, they didn't get nearly to the top of that imposing hill-mass without some initiative and small-unit maneuver. There were no wasteful charges of which Chesty is sometimes accused. A stand-up dash may have been sometimes necessary to exit an ambush zone, but local envelopments and crawling also occurred.

At the Point, 2ndLt. Willis and Cpl. Anderson of 3/1 got close enough across flat unobstructed sand to set the example for all ensuing bunker assaults. White phosphorous and explosives then took the place of the full "blind em, burn em, and blast em" routine.[6] To resolve the impasse at the blockhouse, 1/1's CO used 14-inch shells.[7] So, everyone within the 1st Marine Regiment had been trying to limit casualties.

As in every fight against soldiers from a bottom-up culture, there had been a learning curve for frontline Americans. And young GIs struggling to contend with 3GW-capable adversaries often had good ideas. Some would be tried by buddies, make it into platoon SOPs (standard operating procedures), and occasionally reach a whole battalion. But, not many made it into the tactics manuals after the war. Only through NCO (Non-Commissioned Officer) traditions would a few get shared across the campfires of Korea.

Chesty's Contribution to Progress

By closely studying the battle for Peleliu, modern-day U.S.

warfighters should be able to devise a better way to handle Asian adversaries. There, the Corps most proficient 2GW devotee had faced Japan's new 3GW-oriented defense strategy. With what he and his men learned about that strategy, 1st Marine Division would go on to capture the heart of Okinawa's infamous Shuri Line in 1945 and outsmart several Chinese divisions at the Chosin Reservoir in 1950.

Col. Puller was universally adored by his troops for a number of reasons: (1) never asking them to do anything he wouldn't; (2) not endangering them unnecessarily; (3) frequently interacting with headquarters on their behalf; and (4) truly caring for each and every one of them. After a battle, he would sit down next to the most junior gladiator and ask, "How goes it old man?" For a teenage Gyrene in combat for the first time, such a question from a senior officer served as a testament to his now vital manhood. The same comment helped the parent fireteam and squad to see how important their contribution would be the battle's outcome.

Fireteam Precedents in the Palaus

There's a good way to show how Peleliu's on-the-job training for small units later helped to win WWII in the Pacific. It's something many senior officers still don't realize—the importance of fireteam self-sufficiency. On Iwo, the terrain atop Hill 362-B, at Senda's Last Stand, in Cushman's Pocket, and along the Gorge would be so jumbled that supporting arms couldn't reach its inner recesses, nor could anything bigger than a fireteam.[8] On Okinawa, the frontal slope of Crescent Hill would be too steep and convoluted for artillery bombardment, tanks, or squad-sized bunker assaults.[9]

> Everything about the terrain [on the northern Shuri Line] favored the defenders. The convoluted topography of ridges, draws, and escarpments served to compartment[alize] the battlefield into scores of small firefights, while the general absence of dense vegetation permitted the defenders full observation and interlocking supporting fires from intermediate strongpoints [and the Shuri Heights].[10]
>
> — "The Final Campaign," by Alexander
> History and Museums Division, HQMC, 1996

But, prior to Peleliu, almost all Fleet Marine Corps fireteam activity had been limited to squad participation. In fact, the whole idea of a squad running a full bunker assault—complete with its own base of fire—was revolutionary. Getting the same job done with a fireteam would arise out a need for total surprise.

Events at the Point would provide the initial example of a smaller maneuver element. As with the Marine Raiders at Bairoko, it entailed a doctrinally incorrect double envelopment by buddy teams. Then, hand-delivered smoke came into play. Up to now, smoke grenades had been rarely used because they also blinded the friendly base of fire. But, at the Point, the final pair of Marines couldn't have approached the main gun without them. Something else soon changed in the three-step bunker procedure. Other types of grenades took the place of standoff weaponry. The enemy was both blinded and burned with white phosphorous grenades. Then, a well-directed rifle grenade did the job of hand-delivered satchel charges. That's how trial and error at the Point developed enough tactical technique to take ordinary fireteams to the top of the Umurbrogol.

Which Factors Had Prompted the Improvements

It's no secret that U.S. infantry organizations are too tall and heavily dependent on established procedure. That's why frontline realities have so much trouble reaching the upper echelons (and why Chesty so frequently argued with headquarters). Just in the process of becoming standardized, a good small-unit maneuver can lose much of its vital detail. Staffers must only agree on its basic premise—or "lowest common denominator." Then, they assume each rifleman, buddy team, fireteam, or squad will adapt that denominator to local circumstances. But, 2GW troops are conditioned to do as told. And tired troops with too high a tempo of operations run on muscle memory. So, those missing details or "existing conditions" don't come back into play. Meanwhile, their 3GW-focused opponents have been shown the micro-basics and allowed more movement options.

But, bureaucratic pressures need not erode the effectiveness of every U.S. rifle company. Their leaders have only to ask fireteam members what additional capabilities they would like. Chesty had likely done some of that. Traditionally, the SNCOs (staff NCOs) of each unit learned such things from living in the same "bay" with the troops. When the answers were passed along to the officers, the outfit became less predictable. And, a 2GW-oriented outfit now had frontline fighters with enough skill to try a little 3GW.

Which Small-Unit Stratagems Originated at Peleliu

As *Super-Squad* has demonstrated, the latter stages of WWII produced the modern "state of the art" for small-unit infantry tactics.[11] The arms manufacturers won't agree, but such a conclusion is certainly possible from the intensity of the fighting. In the attempt to defeat a better-armed adversary, the Eastern powers had sought more tactical expertise. Then, having to face German or Japanese squads forced U.S. units to grow.

It's hard to tell how many of the tactical trends from Bairoko, Biak, and Peleliu would help the U.S. military during the rest of the war. This knowledge had been gathered over a period 12 months in three widely separated theaters. That's too well dispersed an instruction to generate many high-level guidelines. And something else took precedence. The U.S. island seizure strategy had overly focused on firepower—to the exclusion of other variables like terrain, weather, and opposition tricks. So while those three battles would lead to a few 3GW methods on Iwo Jima and Okinawa, most would escape the post-war Pentagon. That august agency would go on to suffer a tie in Korea, loss in Vietnam, compromised victory in Iraq, and another loss in Afghanistan. All the while, the U.S. voting public would be told their new war department was doing the best it could under the circumstances. Only to the extent that non-martial aspects of 4th Generation Warfare (4GW) had been assigned to other agencies, would that be true.

Some 2GW Procedures Outlive Their Usefulness

Most tactical researchers (this one included) have been quick to point out the adverse effect of bureaucratic pressure on infantry ability. But, a stiffly hierarchical structure does perform a valuable service at the start of a big war. It enables the military organization to quickly add people with some knowledge of the combat basics. Then, to defeat a thinking adversary, those same people must be allowed to transcend the preliminaries. The resulting increase in frontline initiative then makes possible 3GW.

How This Island Fight Had Made the Pacific War Winnable

The Marines about to invade Peleliu were dealt one of the cruelest "internal blows" imaginable. A senior naval officer suggested the island didn't need to be taken. Veterans like Sledge later talked about being unnecessarily sacrificed.[12] While that sentiment is not new to warfare (many veterans of Vietnam felt the same way), it was not true of either engagement.

At the Great Marianas Turkey Shoot., U.S. forces had just broken the back of Japanese airpower in June 1944. So, for the moment, the Peleliu airfield seemed unnecessary to the invasion of the Southern Philippines. But the *Kamikaze* had yet to appear, and the Leyte landing in October came very close to being a disaster. A huge Japanese fleet came unexpectedly steaming around the corner to confront U.S. transports protected by a few destroyers and "jeep carriers."

> [H]aving passed through the San Bernardino Strait, [Kurita's task force] moved southward along the coast of Samar. By this point, Halsey had shifted the Third Fleet to the north . . . [and] left the American amphibious forces on Leyte woefully unprotected. With the bulk of the Seventh Fleet engaged . . . [elsewhere], all that stood between Kurita and the landing beaches were . . . six escort carriers, three destroyers, and four destroyer escorts.[13]
>
> —*Encyclopedia Britannica.*s.v."Battle of Leyte Gulf"

Had history taken a slightly different turn, the Peleliu airfield would have continued to help the island hopping campaign. And, the removal of a key Japanese hub for reinforcing and resupplying the Southwestern Pacific would have been better appreciated. But, there's another important point to consider.

Japanese ingenuity so far in WWII would manifest itself on Peleliu in September of 1944. That tunnel-laced island was to be the final testing ground for the finishing touches to better way of countering island invasions. Peleliu would be the U.S. Fleet Marine Forces' first exposure to constant ambush and infiltration. Through them, 1st Marine Division discovered how to win the pivotal battle of WWII in the Pacific. It would be Gen. Rupertus' old division finally to capture Shuri Castle (at the heart of Okinawa's final defense line). Some credit for that accomplishment must go to him.

> Daybreak on 29 May 1945 found the 1st Marine Division beginning its fifth consecutive week of frontal assault as part of the U.S. Tenth Army's grinding offensive against the Japanese defenses centered on Shuri Castle in southern Okinawa. Operation ICEBERG, the campaign to seize Okinawa, was now two months old--and badly bogged down. The exhilarating, fast-paced opening of the campaign had been replaced by week after week of costly, exhausting, attrition warfare against the Shuri complex.
>
> The 1st Marine Division, hemmed in between two other divisions with precious little maneuver room, had advanced barely a thousand yards in the past 18 days--an average of 55 yards each bloody day. Their sector featured one bristling, honeycombed ridge line after another--sequentially Kakazu, Dakeshi, and Wana (with its murderous, reverse slope canyon). Just beyond lay the long shoulder of Shuri Ridge, the nerve center of the Japanese Thirty-second Army and the outpost of dozens of the enemy's forward artillery observers who had made life so miserable for American assault forces all month long.
>
> But on this rainy morning, this 29th of May, things seemed some how different, quieter. After days of bitter fighting, American forces had finally overrun both outposts of the Shuri Line: Conical Hill on the east, captured by the 96th Infantry Division, and the Sugar Loaf complex in the west, seized by the 6th Marine Division. Shuri no longer seemed invincible.
>
> Company A of the 1st Battalion, 5th Marines moved out warily, expecting the usual firestorm of Japanese artillery at any moment. There was none. The Marines reached the crest of Shuri Ridge with hardly a firefight. Astonished, the company commander looked westward along the ridge

several hundred yards to the ruins of Shuri Castle, the medieval fortress of the ancient Ryukyuan kings. Everyone in the Tenth Army expected the Japanese to defend Shuri to the death--but the place seemed lightly held. Spiteful small-arms fire appeared to come from nothing more than a rear guard. Field radios buzzed with this astounding news. Shuri Castle itself lay beyond division and corps boundaries, but it was there for the taking. The assault Marines asked permission to seize the prize.

Major General Pedro del Valle, commanding the 1st Marine Division, did not hesitate. By all rights the castle belonged to the neighboring 77th Infantry Division and del Valle knew his counterpart, Army Major General Andrew D. Bruce, would be angry if the Marines snatched the long-sought trophy before his soldiers could arrive. But this was an unprecedented opportunity to grab the Tenth Army's main objective. Del Valle gave the go-ahead. With that, Company A, 1/5, swept west along the ridge against light opposition and took possession of the battered complex. Del Valle's staff had to do some fancy footwork to keep peace with their Army neighbors. Only then did they learn that the 77th Division had scheduled a major bombardment of the castle that morning. Frantic radio calls averted the near-tragedy just in time. Results of the Marines' preemptive action incensed General Bruce. Recalled del Valle: "I don't think a single Army division commander would talk to me after that."

Notwithstanding this inter-service aggravation, the Americans had achieved much this morning. For two months the Shuri Heights had provided the Japanese with superb fields of observed fire that covered the port city of Naha and the entire five-mile neck of southern Okinawa. Even now, as the Marines of A/1/5 deployed into a hasty defensive line within the Castle's rubble, they were oblivious to the fact that a Japanese rear guard still occupied portions of the mammoth underground headquarters complex directly under their muddy boondockers. They would be astounded to learn that the subterranean headquarters of the Thirty-Second Army measured 1,287 feet long and as much as 160 feet deep--all of it dug by pick and shovel.

The Japanese had in fact stolen a march on the approaching Tenth Army. Most of their forces had retreated southwards during the incessant rains, and would soon occupy the third (and final) ring of their prepared, underground defenses, a series of fortified escarpments in the Kiyamu Peninsula.

Seizing Shuri Castle represented an undeniable milestone in the Okinawa campaign, but it was a hollow victory. Just as the flag-raising on Iwo Jima's Mount Suribachi signified only the end of the beginning of that prolonged battle, the capture of Shuri did not end the fighting. The brutal slugfest on Okinawa still had another 24 days to run. And still the Plum Rains fell, and the horrors, and the dying, continued.[14]

— *The Final Campaign,* by Alexander
History and Museums Division, HQMC, 1996

In Humble Retrospect

Prior to Peleliu, the worldwide refinements to squad maneuver had escaped the average Leatherneck. Some of the moves had existed in the Orient for centuries. Others had been developed during the latter stages of WWI. And still others arose out of the extreme savagery of WWII Europe. But, to avoid the bloodshed of close combat, most USMC leaders preferred standoff firepower. Some of the Bairoko lessons would reach the Iwo outfits in February 1945 through former Raiders, and Peleliu tutorials be carried to Okinawa by 1st Division. Without them, neither island would have been as easily captured. In short, those tactical innovations allowed America to win the Pacific War against a foe that was more maneuver oriented. Japan's large-unit pushes now followed a sharply pointed "spear"—infantry squads tactically adroit enough to totally surprise the quarry.

Whether or not Peleliu was a necessary stepping stone to the Japanese homeland, its denial to the enemy as a forward supply, training, and staging point for the Dutch East Indies and New Guinea was still important.[15] So, the honored war dead of Peleliu can rest in peace, knowing that their sacrifice did make a major difference. In fact, what they learned the hard way on Peleliu may now permit modern-day GIs to blockade the Straits of Malacca with minimal loss of life. No amount of technology can replace what needs to happen at the fireteam level in that kind of combat. And any invasion target not blanketed with tiny outposts during its consolidation phase will risk enemy takeover. This will be doubly true where the whole place sits atop a labyrinth of underground rooms and passageways.

The Chinese Peleliu of the Future

U.S. technology has yet to determine from the air what lies below ground. No amount of non-nuclear explosive will destroy a sub-surface installation that has not been precisely located. So, the Pentagon still lacks the capacity to invade a place like Peleliu with minimal loss of life.

If Peleliu had been secretly fortified since 1920, then other strategically vital places within Japan's "Greater Asian Co-Prosperity Sphere" were as well. That would include Great Natuna Island and the Anambas at the Indonesian entrance to the Straits of Malacca.[16] By the end of the Pacific War, the IJA had even burrowed through the walls of the Hue City Citadel. That's why that fortress was so hard to take in 1968.[17]

> The Citadel wall . . . [was] as many as seventy-five meters thick in places and honeycombed throughout by passages and bunkers excavated by the Japanese near the end of World War II.[18]
>
> — *Fire in the Streets,* by Hammel

Great Natuna or Other PRC Outposts Will Have Good MLRs

The Peleliu struggle provided a new axiom of war that was confirmed on Iwo Jima and Okinawa. Because of the sub-surface nature of a well-prepared MLR, there must be an initial effort to spot/silence some of its guns before any deep penetration is attempted. On Iwo Jima, that effort came from machinegun-enveloping scouts.[19] On Okinawa, it was provided by forward-ranging fireteams and squads during lulls in the U.S. shelling.

> At first, the Americans [on Okinawa] did not know the pillbox caves even existed and had no doctrine for dealing

> with them. . . . The Americans at the outset of the offensive advanced *en masse* and were mowed down in crowds by cave-hidden machine guns when they reached the reverse slopes (Inagaki, *Okinawa,* pp. 139-140). American officers on the scene quickly developed a method for avoiding these unhappy effects, however. They would bombard the Japanese positions they faced, forcing the IJA off the surface. They would then infiltrate men in small numbers through narrow gaps in the Japanese fixed-cave fire line ("dead spaces").[20]
>
> – *Leavenworth Papers No. 18,* 1990

In other words, bad things will happen if the targeted sector is not initially "inspected and partially cleared" by U.S. infiltrators. That MLR preparation can best be accomplished by tiny teams of truly light (GW trained and non-vehicular) riflemen. Peleliu's precipitous and below-ground dimensions would more greatly challenge the fully equipped American infantrymen of today than it did the low-tech ground pounders of 1944. Nor would modern-day U.S. special operators have enough assault expertise to be of much help.

Never forget these gifts of tactical insight from the fine young Americans on Peleliu. They gave it their all so that others could gain the ultimate victory.

> "Greater love hath no man than . . . [he] lay down his life for his friends."[21]
>
> – *Christian Bible,* King James Version

Notes

SOURCE NOTES

Illustrations:

Cover art and images on pages 189 and 227 are from "The 2000 Yard Stare," by Thomas Lea. Retrieved in August 2020 from WIKIMEDIA COMMONS, where it's attributed to "U.S. Army, Tom Lea/Public domain." While the rights to this painting are owned by the creator's estate, almost all content hosted by Wikimedia Commons may be freely reused subject to the following disclosures: (1) attribution of the original creator; and (2) identification of its license. In this case. the latter was the Creative Commons Attribution-ShareAlike 3.0 License and the GNU Free Documentation License. While Thomas Lea was a Life Magazine employee when he made this painting, its reprinting by an educational nonprofit constitutes fair usage because it appeared on the cover of a Marine Corps publication—"Bloody Beaches: The Marines at Peleliu" by Gayle. At Wikimedia Commons, it has image designator, "Tom_Lea_-_2000_Yard_Stare.jpg." While the central figure is supposedly a Marine with battle fatigue, the self-propelled artillery piece in the background belongs to the U.S. Army. 1st Marine Division's landing force reserve—the 81st Infantry Division—had helped with the latter stages of the battle. This painting is now part of the U.S. Army Center of Military History collection at Fort Lesley J. McNair, Washington, D.C.

Image on page 3 is part of Stefan R. Landsberger Collection at Leiden University in Holland under the care of THE INTERNATIONAL INSTITUTE OF SOCIAL HISTORY (IISH). Chinaposters.net explains the reprint rights. Designers/artists of such posters were employees of art academies, museums, or publishers. During socialism's high tide, it was counterrevolutionary to personally claim any copyrights. Such institutions would have been the copyright holders. According to IISH, none exercised any copyrights for materials published from 1949 to the mid-1980's. Until then, China was not a signatory of any International Copyright Convention. Thus, IISH thinks China will find it hard to claim copyrights retroactively. In principle, IISH sees no problem with Posterity Press reprinting this image for educational purposes.

Pictures on pages 23 and 210 are from NATIONAL GUARD IMAGE GALLERY. Entitled "Jungleers on Biak" and "Red Arrow at War," they were painted by Keith Rocco and Michael Gnatek, respectively. They were retrieved in September 2020 from the following url: https://www.nationalguard.mil/Resources/Image-Gallery/Historical-Paintings/Heritage-Series/. Only for an educational nonprofit is their reprinting fair usage. Others must be more careful of a possible copyright.

Map on pages 24 is in the public domain from WIKIMEDIA COMMONS. It comes from "Reports of General MacArthur," in *The Campaigns of MacArthur in the Pacific,* volume 1, U.S. Army Center of Military History. It was retrieved in September 2020 from the following url: https://commons.wikimedia.org/wiki/File:Biak_Island_Operation_May-August_1944.jpg.

Map on page 51 was reprinted with permission of Ken Feisel at "ARCHEOLOGY MAGAZINE." It was retrieved in September 2020 from the following url: https://www.archaeology.org/issues/344-1907/features/7712-peleliu-wwii-pacific-battle.

Picture on page 52 is in the public domain from WIKIMEDIA COMMONS. Entitled "Palau under German rule," it was painted by Rudolf Hellgrewe around 1908.

Maps on page 60 and 107 were reprinted after asking permission of HISTORYOFWAR.ORG. It is a "Japanese Map of Peleliu" possibly from captured 31st Army files. It was retrieved in August 2020 from the following urls: http://www.historyofwar.org/Pictures/pictures_peleliu12.html. As an educational nonprofit, Posterity Press considers the reprinting of this low resolution image to be fair usage. Others must be more careful of a copyright.

Images on pages 65, 97, 117, 135, 151, 169, 183, 191, and 203 were available under the fair usage provision because they come from a U.S. MARINE CORPS-produced pamphlet—"Bloody Beaches: The Marines at Peleliu" by Gayle. While Thomas Lea was a *Life Magazine* employee when he made these drawings in 1944 at Peleliu, the reproduction by an educational nonprofit of these low-resolution extracts from that pamphlet now constitutes fair usage.

Image on page 69 is from the NATIONAL MUSEUM OF THE MARINE CORPS, Quantico, VA, on 15 Jan 2007. With image designator "d83099f461350967f00fea1a82d43fbb.jpg." it shows the multi-color cover artwork of a 1st Marine Division cruise book that had been apparently based on a black and white photograph of the landing. The original painting is likely in the public domain. If not, its reprinting constitutes fair usage for an educational nonprofit.

Map on page 71 reprinted with permission of MERIDIAN MAPPING, Minneapolis, MN. It comes from "Storming of the Point at Peleliu" at the *Warfare History Network*, n.d. With image designator, "WQ-Peleliu-MAP-2-4C-Sum10.jpg," it was retrieved in August 2020 from the following url: https://warfarehistorynetwork.com/2018/12/23/storming-the-point-at-peleliu/?unapproved=10393&moderation-hash=b055ef0f2707ce7576d5db3d4ea2d642#comment-10393.

Picture on pages 84 reprinted from "Artphoto Archives," U.S. ARMY'S CENTER OF MILITARY HISTORY. Appearing to have been painted by an active-duty U.S. Army member, it is considered to be in the public domain, with the following image designator: "avop05-99_1.jpg".

Image on page 89 is from UNIVERSITY OF MISSOURI AT ROLLA. It comes from "Japanese Defenses and Fortifications Tawara, Iwo, Jima, and Okinawa 1943-1945," by J. David Rogers and Karl F. Hasserman Chair, n.d. As this study contains no copyright notification, the reprint of this picture by an educational nonprofit is assumed to be fair usage.

Maps on pages 114, 175, 180, and 205 come from "Preliminary Draft—1/91 (of Peleliu study)," BOTANY DEPARTMENT., UNIVERSITY OF HAWAII, January 1991. This study was done for the Historic Preservation Office, Trust Territory of the Pacific Islands. Their reprinting by an educational nonprofit is considered fair usage. But, others may have to be more wary of a copyright.

Picture on page 121 reproduced after written assurance from Cassell PLC, London, that the copyright holder for *WORLD ARMY UNIFORMS SINCE 1939,* text by Andrew Mollo and Digby Smith, color plates by Malcolm McGregor and Michael Chappell, could no longer be contacted. The illustration is from Part I (Plate 156).

Maps on page 140 and 174 reprinted with permission of NAVAL HISTORY MAGAZINE. With image designator, "PeliliuLanding SO 19. png," it's from an article entitled "Eight Days in Hell" in the October 2019 issue. It was retrieved in September 2020 from this url: https://www.usni.org/magazines/naval-history-magazine/2019/october/eight-days-hell.

Maps on pages 178 and 179 come from University of Texas at Austin, PERRY CASTANEDA MAP COLLECTION. Palau Islands 1:12,500. Series W854, U.S. Army Map Service, 1946. Ngarmoked NW-D. Sheet 7331 I NW. Public domain. As retrieved in December 2020 from the following url: http://legacy.lib.utexas.edu/maps/ams/palau/txu-oclc-6573573-7331-1-nwd.jpg.

Picture on page 215 is entitled "Planting of 20K Flags on Memorial Day," by Ellen White. It comes from the U.S. ARMY'S CENTER OF MILITARY HISTORY. Its reprinting by an educational nonprofit is considered to be fair usage, but other must be more mindful of a copyright. Copyright © n.d. All rights reserved.

Text:

Excerpts from *THE PACIFIC WAR ONLINE ENCYCLOPEDIA* reprinted with permission of its website manager, trollingshallows@msn.com. All rights reserved. Copyright © n.d.

Excerpts reprinted after asking permission of HISTORYOFWAR.ORG. Its from a publication by Peter Antill in 2003 entitled "Peleliu, battle for (Operation Stalemate II) - The Pacific War's Forgotten Battle, September-November 1944." It was retrieved in January 2023 from the following url: http://www.historyofwar.org/articles/battles_peleliu.html#pocket. As an educational nonprofit, Posterity Press considers the reprinting of this excerpt to be fair usage. Others must be more mindful of a copyright. Copyright © n.d. All rights reserved.

ENDNOTES

Chapter 1: *The Chinese Communist Co-Prosperity Sphere*

1. *Wikipedia Encyclopedia,* s.v. "Greater East Asia Co-Prosperity Sphere."
2. "China's Military Rise: The Dragon's New Teeth," *The Economist,* 7 April 2012.
3. "How Uninhabited Islands Soured Japan-China Ties," *BBC News* on line, 10 November 2014.
4. "China's Newest Maritime Dispute," by Zachary Keck, *The Diplomat* on line, 20 March 2014.
5. *Wikipedia Encyclopedia,* s.v. "Territorial Disputes in the South China Sea."
6. Robert A. Manning, *The Asia Energy Factor* (New York: Polygrave, 2000), in *China: The Gathering Threat,* by Menges (Nashville, TN: Nelson Current, 2005), p. 379; Bill Gertz, *The China Threat: How the People's Republic Targets America* (N.p.: Regnery Publishing, 2000), p. 187.
7. David G. Winneck, "The South China Sea Dispute Background Briefing," manuscript, 11 February 1999, in *China: The Gathering Threat,* by Menges (Nashville, TN: Nelson Current, 2005), pp. 294-297.
8. Maria rost Rublee, "Foreign Policy Responses to Aggressive Territorial Moves: The Case of the Spratly Islands," *Internat. Affairs Review,* Summer 1996, in *China: The Gathering Threat,* by Menges (Nashville, TN: Nelson Current, 2005), p. 295.
9. David G. Winneck, "The South China Sea Dispute Background Briefing," manuscript, 11 February 1999, in *China: The Gathering Threat,* by Menges (Nashville, TN: Nelson Current, 2005), pp. 294-297.
10. Ibid.; Richard Fisher, *PLA Modernization Gains Momentum* (N.p.: Heritage Foundation, 1997), p. 2, in *China: The Gathering Threat,* by Menges (Nashville, TN: Nelson Current, 2005), pp. 294-297.
11. Menges, *China: The Gathering Threat (*Nashville, TN: Nelson Current, 2005), p. 295.
12. Bill Gertz, "China Makes Upgrades to Island Base," *Washington Times,* 11 February 1999, p. A12, in *China: The Gathering Threat,* by Menges, pp. 294-297.
13. Ibid.
14. "Two U.S. carrier groups conduct exercises in South China Sea," by Se Young Lee, *Reuters* on line, 9 February 2021; "U.S. Warship Sails Near Chinese-Controlled S. China Sea Islands," by Reuters Wire Service, *U.S. News & World Report* on line, 5 February 2021.

15. "U.S. Navy Sends Destroyer by Man-Made Islands in Challenge to China's Claims," by David Larter, *Navy Times* on line, 27 October 2015.

16. "China Stations Combat Aircraft on South China Sea Island," by Franz-Stefan Gady, *The Diplomat* on line, 10 November 2015.

17. *NPR's "Morning Edition" News,* 17 February 2016.

18. Ibid., 1 June 2015; "China Defends Land Reclamation in the Spratly Islands," *Military Times* on line, 24 November 2014.

19. "Up to the Task," by Audrey McAvoy (AP), *Jacksonville Daily News* (NC), 17 January 2016, p. A5.

20. "Report: China Has Reclaimed 3,200 Acres in South China Sea," by Lolita Baldor (AP), *MSN News* on line, 14 May 2016.

21. "Inside the International Contest over the Most Important Waterway . . . ," by Hannah Beech, *Time Magazine* on line, 6 June 2016, pp. 42-51.

22. Ibid.

23. Ibid.

24. "The Meaning of China's Defeat in the South China Sea," by Hannah Beech, *Time Magazine* on line, 25 July 2016, p. 12.

25. *Wikipedia Encyclopedia,* s.v. "Scarborough Shoal."

26. "How the Scarborough Shoal came back to haunt China-Philippines relations," by Richard Heydarian, *South China Morning Post* on line, 23 June 2018.

27. *NPR's "Morning Edition News,"* 5 September 2016.

28. "China to soon build air, naval bases in Scarborough Shoal, Carpio warns," by CNN Staff, *CNN Philippines* website, 9 June 2020.

29. *Wikipedia Encyclopedia,* s.v. "Greater East Asia Co-Prosperity Sphere," Wikimedia Commons global image (colored portions).

30. "Military Build-Up in the South China Sea," by Eric Grossman, Rand Corp. website, n.d.

Chapter 2: *The Bairoko Incident*

1. *Marines in the Central Solomons,* by Maj. John N. Rentz, USMC Historical Monograph (Washington, D.C.: Hist. Branch, HQMC, 1952), chapt. 4—From Rice to Bairoko.

2. Ibid.

3. Assoc. of U.S. Marine Raiders website, s.v. "Chronology."

4. "From Makin to Bougainville: Marine Raiders in the Pacific War," by Major Jon T. Hoffman, *Marines in WWII Commemorative Series* (Washington, D.C.: Marine Corps Hist. Ctr., 1995), p. 34.

5. *Gung Ho: The Corp's Most Progressive Tradition,* by H. John Poole (Emerald Isle, NC: Posterity Press, 2012), glossary.

6. Ibid., pp. 74-79.
7. Ibid., chapt. 10.
8. "Isolation of Rabaul," by Shaw and Kane, *History of U.S. Marine Corps Operations in World War II Series* (Washington, D.C.: Historical Branch, HQMC, 1963), vol. II, chapt. 4—The Dragon's Peninsula Campaign, pp. 130-145.
9. "From Makin to Bougainville," by Hoffman, p. 33.
10. *Marines in the Central Solomons,* by Rentz, "USMC Historical Monograph," chapt. 4.
11. "Isolation of Rabaul," by Shaw and Kane, pp. 137, 138.
12. *Iwo Jima: Amphibious Epic,* by Lt.Col. Whitman S. Bartley (Washington, D.C.: Hist. Branch, HQMC, 1954), p. 140.
13. *Marines in the Central Solomons,* by Rentz, "USMC Historical Monograph," chapt. 4, pp. 113, 114.
14. Bruce I. Gudmundsson, *Stormtroop Tactics—Innovation in the German Army 1914-1918* (New York: Praeger, 1989), pp. 43-50, 94, 146, 147.
15. *Gung Ho,* by Poole, chapt. 10.
16. *North Korea's Hidden Assets,* by H. John Poole (Emerald Isle, NC: Posterity Press, 2018), chapt. 2.
17. "From Makin to Bougainville," by Hoffman, p. 34.
18. "Isolation of Rabaul," by Shaw and Kane, pp. 130-145.
19. Ibid.
20. "From Makin to Bougainville," by Hoffman, p. 33.
21. "Nomonhan: Japanese—Soviet Tactical Combat, 1939," by Edward J. Drea, *Leavenworth Papers No. 2* (Ft. Leavenworth, KS: Combat Studies Inst., U.S. Army Cmd. & Gen. Staff College, 1981), p. 62.
22. *Marines in the Central Solomons,* by Rentz, "USMC Historical Monograph," chapt. 4.
23. "Isolation of Rabaul," by Shaw and Kane, p. 138.
24. "From Makin to Bougainville," by Hoffman, pp. 35, 40.
25. Ibid., p. 35.

Chapter 3: *The Assault on Biak*

1. "History of the Biak Operation 15-27 June 1944," 41st Division, U.S. Army, 1944, Armed Forces Staff College Library copy, p. 1.
2. "The Approach to the Philippines," by Robert Ross Smith, *U.S. Army in WWII Series* (Washington, D.C.: Ctr. of Mil. Hist., 1996), p. 334.
3. "History of the Biak Operation 15-27 June 1944," 41st Division, p. 2.
4. "The Approach to the Philippines," by Smith, p. 280.

5. *Super-Squad: The Now Missing Component,* by H. John Poole (Emerald Isle, NC: Posterity Press, January 2020). chapt. 14; *Isolation of Rabaul,* by Shaw and Kane, part II, chapt. 4—The Dragon's Peninsula Campaign, p. 133.

6. "From Makin to Bougainville," by Hoffman, p. 33; *Marines in the Central Solomons,* by Rentz, "USMC Historical Monograph," chapt. 4, p. 112.

7. *Wikipedia Encyclopedia,* s.v. "Battle of Biak."

8. Joseph C. Goulden, *Korea: The Untold Story of the War* (New York: Times Books, 1982), p. 295.

9. "The Approach to the Philippines," by Smith, p. 306.

10. "The Battle of Biak: 'A Terrifying Glimpse Into the Soul of Mankind," by David Alan Johnson, *Warfare History Network,* 30 December 2018.

11. Ibid.

12. Ibid.

13. Ibid.

14. Ibid.

15. Ibid.

16. Ibid.

17. Ibid.

18. Ibid.

19. Ibid.

20. Ibid.

21. Ibid.; "The Approach to the Philippines," by Smith, pp. 391, 392, and footnotes 41, 47.

22. "The Approach to the Philippines," by Smith, p. 302.

23. "History of the Biak Operation 15-27 June 1944," 41st Division, p. 43.

24. "The Approach to the Philippines," by Smith, p. 366.

25. Ibid., p. 389.

26. Ibid., p. 301.

27. "The Battle of Biak: 'A Terrifying Glimpse Into the Soul of Mankind'," by Johnson.

28. "The Approach to the Philippines," by Smith, p. 366.

29. "The Battle of Biak: 'A Terrifying Glimpse Into the Soul of Mankind'," by Johnson.

30. "History of the Biak Operation 15-27 June 1944," 41st Division, p. 42.

31. *The Iwo Alamo,* by H. John Poole (Emerald Isle, NC: Posterity Press, 2020), fig. 3-1.

32. "The Approach to the Philippines," by Smith, p. 301.

33. "History of the Biak Operation 15-27 June 1944," 41st Division, p. 15.

34. "The Approach to the Philippines," by Smith, p. 301.

35. *The Pacific War Online Encyclopedia*, s.v. "The Battle of Biak."
36. "The Approach to the Philippines," by Smith, p. 389; "History of the Biak Operation 15-27 June 1944," 41st Division, p. 46.
37. "The Battle of Biak: 'A Terrifying Glimpse Into the Soul of Mankind'," by Johnson.
38. "History of the Biak Operation 15-27 June 1944," 41st Division, p. 31.
39. "The Approach to the Philippines," Smith, pp. 388, 389.
40. *Marines in the Central Solomons,* by Rentz, "USMC Historical Monograph," chapt. 4, p. 112.
41. "History of the Biak Operation 15-27 June 1944," 41st Division, p. 9.
42. Ibid., p. 16.
43. Ibid.
44. *Super-Squad,* by Poole, chapt. 14.
45. *North Korea Handbook,* PC-2600-6421-94 (Washington, D.C.: DoD, 1994), p. 3-99.
46. "The Battle of Biak: 'A Terrifying Glimpse Into the Soul of Mankind'," by Johnson.
47. *The Tiger's Way: A U.S. Private's Best Chance at Survival,* by H. John Poole (Emerald Isle, NC: Posterity Press, 2003), p. 179.
48. Unidentified NCOs to Chesty Puller, in "Fighting on Guadalcanal," U.S.A. War Office pamphlet (Washington, D.C.: U.S. Govt. Printing Office, 1942), pp. 35, 37.
49. Gudmundsson, *Stormtroop Tactics*, pp. 43-50, 94, 146, 147.
50. Memorandum for the record, by H.J. Poole, while a rifle company commander for eight months south of DaNang.
51. *Gung Ho,* by Poole, chapt. 10.
52. Ibid.

Chapter 4: *Next Stepping Stone for U.S. in the Pacific*

1. "Strategic and Operational Importance of Peleliu During the Pacific War," by Maj. John L Medeiros, masters thesis, Cmd. & Staff College, Marine Corps Univ., 2010, p. 4.
2. Jim Moran and Gordon Rottman, *Peleliu 1944: The Forgotten Corner of Hell* [Oxford: Osprey, 2002], p. 12, in "Strategic and Operational Importance of Peleliu . . . ," by Medeiros, p. 5.
3. Ibid.
4. "Strategic and Operational Importance of Peleliu . . . ," by Medeiros, p. 11.
5. Ibid., p. 12; *The Assault on Peleliu,* by Maj. Frank R. Hough USMCR [Washington, D.C. Historical Div., HQMC, 1950), p. 135.

6. *The Assault on Peleliu,* by Maj. Frank R. Hough USMCR [Washington, D.C. Historical Div., HQMC, 1950), p. 194.

7. *The War,* by Ken Burns (N.p.: Sony Legacy, 2007), *PBS,* 12 September 2020.

8. *The Assault on Peleliu,* by Hough, pp. 1-4.

9. "Eight Days in Hell," by Colonel Richard D. Camp USMC (Ret.), *Naval History Magazine,* vol. 34, no. 5, October 2019; *The Assault on Peleliu,* by Hough, p. 30.

10. "WWII Battlefield Survey of Peleliu Island Peleliu State, Republic of Palau," by Rick Knecht, Neil Price, and Gavin Lindsay (Washington, D.C.: Nat. Park Service American Battlefield Protection Program, 2012), p. 23.

11. John Shaw and the editors of Time-Life Books, *Red Army Resurgent* (Chicago: Time-Life Books, 1979), pp. 163-165.

12. "Reflections on Iwo Jima: The Japanese Tunnels," interview of Col. Dave Severance USMC (Ret.)., in *The Iwo Alamo,* by Poole, p. 241.

13. Philip Warner, *Japanese Army of World War II,* Men-at-Arms Series, color plates by Michael Youens (London: Osprey Publications Ltd., 1972), pp. 23-25.

14. President John F. Kennedy, quote from his 1962 West Point graduation speech, in "Unconventional Warfare Fundamentals," at *The Irregular Warrior* website.

15. *Wikipedia Encyclopedia,* s.v. "Battle of Peleliu."

16. *The Assault on Peleliu,* by Hough, p. 192.

17. "Japan WW2 soldier who refused to surrender Hiroo Onoda dies," *BBC News* on line, 17 January 2014.

18. *The Assault on Peleliu,* by Hough, p. 183.

19. Ibid., p. 18.

20. "National Register of Historic Places Inventory—Nomination Form (for Peleliu Battlefield, Palau Islands)," National Park Service, U.S. Dept. of the Interior, n.d., item 8, p. 2; "Place of the Loyal Samurai," by Daniel Weiss, *Archaeology Magazine*, July/August 2019; "WWII Battlefield Survey of Peleliu Island . . .," by Knecht, Price, and Lindsay, p. 8.

21. "Peleliu 1944: The Archaeology of a South Pacific D-Day," by Neil Price and Rick Knecht, Dept. of Archaeology, Univ. of Aberdeen, England, *Journal of Conflict Archaeology,* vol. 7, no. 1, January, 2012, p. 7; *The Assault on Peleliu,* by Hough, p. 17.

22. *The War,* by Burns.

23. Jeffrey M. Moore, *Spies for Nimitz: Joint Military Intelligence in the Pacific War* [Annapolis, MD. Naval Inst. Press. 2004], p. 150, in "Strategic and Operational Importance of Peleliu . . . ," by Medeiros, pp. 19, 20.

24. Eugene Sledge [Peleliu veteran] as quoted, in *The War*, by Burns.

25. *The Assault on Peleliu,* by Hough, p. 183.

Chapter 5: *The Initial Ambush at Peleliu*

1. "Eight Days in Hell," by Camp.
2. *With the Old Breed: At Peleliu and Okinawa,* by E.B. Sledge (New York: Presidio Press, div. of Random House, 2007), p. 71; "Battle of Peleliu Facts," by Mack Dean, *WorldWar2Facts,* 2 June 2020; "Battle of Peleliu" *Facts and Details,* n.d.; *Bloody Beaches: The Marines at Peleliu,* by B.Gen. Gordon D. Gayle USMC (Ret.), *Marines in WWII Commemorative Series* (Washington, D.C.: Marine Corps Historical Ctr., 1996), p. 18.
3. *The Iwo Alamo,* by Poole, chapt. 1, endnotes 12, 19.
4. Ibid., p. 124.
5. *With the Old Breed,* by Sledge, pp. 62. 63.
6. "Storming the Point at Peleliu," by Dick Camp, *Warfare History Network,* 23 December 2018.
7. "The Battle of Peleliu: Shocked Beyond Imagination," by Al Hemingway, *Warfare History Network,* 4 January 2019:*The Assault on Peleliu,* by Hough, p. 37, and *Japanese CenPac Ops*, p. 84, in "Western Pacific Operations," *History of Marine Corps Operations in World War II,* vol. IV. part III, by George W. Garand and Truman R. Strobridge (N.p.: Historical Branch, G-3 Div., HQMC 1971), p. 109; "Battle of Peleliu Facts," by Mack Dean, *WorldWar2Facts,* 2 June 2020; "Unnecessary Hell: The Battle of Peleliu," by Eric Niderost, *Warfare History Network,* 10 August 2016; "National Register of Historic Places Inventory—Nomination Form," item 8, p. 3.
8. "Storming the Point at Peleliu," by Camp.
9. Ibid.
10. *Bloody Beaches: The Marines at Peleliu,* by B.Gen. Gordon D. Gayle USMC (Ret.), *Marines in WWII Commemorative Series* (Washington, D.C.: Marine Corps Historical Ctr., 1996), pp. 4. 5.
11. "The Approach to the Philippines," by Smith, p. 497; *Bloody Beaches,* by Gayle, small map.
12. "Storming the Point at Peleliu," by Camp.
13. *Home of the Heroes,* Navy Cross citation for Sgt. Steve Gerycz.
14. "Western Pacific Operations," *History of Marine Corps Operations in World War II,* vol. IV. part III, by George W. Garand and Truman R. Strobridge (N.p.: Historical Branch, G-3 Div., HQMC 1971), pp. 114, 115.
15. *The Assault on Peleliu,* by Hough, p. 17; "Japanese Map of Peleliu," possibly one of Eugene Sledge's non-published finds, retrieved in August 2020 from the following url: http://www.historyofwar.org/Pictures/pictures_peleliu12.html.

16. *Bloody Beaches,* by Gayle, pp. 6, 7.
17. Ibid., pp, 11-14; "Western Pacific Operations," by Garand and Strobridge, p. 73.
18. *Bloody Beaches,* by Gayle, pp. 11-13.
19. Ibid., p. 6.
20. "Eight Days in Hell," by Camp.
21. *Home of the Heroes,* Navy Cross citation for Cpl. Herbert B. Goff, Jr.
22. *Bloody Beaches,* by Gayle, p. 6.
23. 1/1 Afteraction Report, in "Eight Days in Hell," by Camp.
24. Ibid.
25. "Eight Days in Hell," by Camp.
26. "The Approach to the Philippines," by Smith, p. 412; *Chesty: The Story of Lieutenant General Lewis B. Puller,* by Col. Jon T. Hoffman USMCR (New York: Random House, 2002), p. 273; "Storming the Point at Peleliu," by Camp, up-close picture of Point bunker's rear entrance.
27. *Chesty,* by Hoffman, p. 269; "Western Pacific Operations," by Garand and Strobridge. p. 79.
28. "Storming the Point at Peleliu," by Camp.
29. *Chesty,* by Hoffman, p. 273.
30. Ibid.
31. "Storming the Point at Peleliu," by Camp.
32. *Home of the Heroes,* Navy Cross citation for Sgt. John Koval.
33. *Marines in the Central Solomons,* by Rentz, "USMC Historical Monograph," chapt. 4.
34. "Storming the Point at Peleliu," by Camp; *The Assault on Peleliu,* by Hough, p. 40.
35. *The Assault on Peleliu,* by Hough, p. 41, footnote 15.
36. *Wikipedia Encyclopedia,* s.v. "spider hole."
37. Col. Hiromichi Yahara, *The Battle for Okinawa,* trans. Roger Pineau and Masatoshi Uehara (New York: John Wiley & Sons, 1995), p. 228.
38. "History of the Biak Operation 15-27 June 1944," 41st Division, p. 9.
39. *Home of the Heroes,* Navy Cross citation for Cpl. Henry W. Hahn.
40. *Chesty,* by Hoffman, p. 273.
41. *Home of the Heroes,* Navy Cross citation for Capt. George P. Hunt.
42. *Sands of Iwo Jima* (Hollywood: Republic Pictures, 1988), videocassette.
43. *Home of the Heroes*, Navy Cross citation for 2ndLt. William L. Willis.

44. "Western Pacific Operations," by Garand and Strobridge, p. 112.
45. Ibid., p. 41.
46. "Storming thc Point at Pclcliu," by Camp.
47. *Home of the Heroes,* Navy Cross citation for PFC Victor L. Case.
48. Ibid., Navy Cross citation for 2ndLt. Lewis Cladin.
49. *Gung Ho,* by Poole, chapt. 10.
50. Ibid., chapt. 11.
51. *Bloody Beaches,* by Gayle, p. 6.
52. "Eight Days in Hell," by Camp.
53. *Bloody Beaches,* by Gayle, p. 6.

Chapter 6: *Peleliu's Developing Ambush Pattern*

1. "Eight Days in Hell," by Camp.
2. *Phantom Soldier: The Enemy's Answer to U.S. Firepower,* by H. John Poole (Emerald Isle, NC: Posterity Press, 2001), chapt. 12.
3. *The Iwo Alamo,* by Poole, chapt. 11.
4. Western Pacific Operations," by Garand and Strobridge, p. 73.
5. *Bloody Beaches,* by Gayle, p. 6.
6. *Home of the Heroes,* Navy Cross citation for Sgt. Charles R. Monarch.
7. *Bloody Beaches,* by Gayle, p. 10.
8. "Eight Days in Hell," by Camp.
9. *Wikipedia Encyclopedia,* s.v. ".50 BMG" and "Madsen 20mm Cannon."
10. "National Register of Historic Places Inventory—Nomination Form," item 8, p. 2.
11. "Eight Days in Hell," by Camp.
12. *With the Old Breed,* by Sledge, p. 69.
13. "Japanese Defenses and Fortifications Tawara, Iwo, Jima, and Okinawa 1943-1945," by J. David Rogers and Karl F. Hasselmann (Univ. of Missouri at Rolla, n.d.), p. 53.
14. *Super-Squad,* by Poole, p. 21.

Chapter 7: *Extent of Subterranean Chambers on Peleliu*

1. "Japanese Military Caves on Peleliu," *Know Your Enemy,* CinCPac-CinCP0A Bulletin 173-45, in *The Assault on Peleliu,* by Hough, Appendix C.
2. *The Assault on Peleliu,* by Hough, p. 7.
3. Ibid., p. 9.
4. "WWII Battlefield Survey of Peleliu Island . . . ," by Knecht, Price, and Lindsay, p. 263.

5. "Storming the Point at Peleliu," by Camp.

6. "The Battle of Peleliu: Shocked Beyond Imagination," by Hemingway.

7. *The Assault on Peleliu,* by Hough, pp. 193-196.

8. Ibid., appendix D.

9. "Peleliu 1944: The Archaeology of a South Pacific D-Day," Price and Knecht, note 5; "The Battle of Peleliu: Shocked Beyond Imagination," by Hemingway; "WWII Battlefield Survey of Peleliu Island . . .", by Knecht, Price, and Lindsay, pp. 44, 154, 182, 183, 226, 232.

10. *The Assault on Peleliu,* by Hough, pp. 37, 39, 59, 137, 138, 180.

11. Ibid., p. 42.

12. *Bloody Beaches,* by Gayle, p. 6.

13. *The Assault on Peleliu,* by Hough, p. 210.

14. Ibid., p. 18.

15. *Bloody Beaches,* by Gayle, p. 28; Denfeld, as quoted in "National Register of Historic Places Inventory—Nomination Form," item 7, p. 4.

16. *The Assault on Peleliu,* by Hough, pp. 193-196, and Map 3.

17. JANIS no. 103 vol. I, p. VIII, 4, in *The Assault on Peleliu,* by Hough, appendix D.

18. *The Assault on Peleliu,* by Hough, pp. 193-196.

19. "Eight Days in Hell," by Camp.

20. *The Assault on Peleliu,* by Hough, p. 138.

21. "Eight Days in Hell," by Camp.

22. *The Assault on Peleliu,* by Hough, p. 42.

23. *Bloody Beaches,* by Gayle, p. 20.

24. Ibid., p. 25.

25. *The Assault on Peleliu,* by Hough, p. 181.

26. "Western Pacific Operations," by Garand and Strobridge, p. 110; 1/1/ CO, in "Eight Days in Hell," by Camp.

27. *Bloody Beaches,* by Gayle, p. 6.

28. Western Pacific Operations," by Garand and Strobridge, p. 116; "National Register of Historic Places Inventory—Nomination Form," item 8, p. 4.

29. *Bloody Beaches,* by Gayle, p. 16.

30. "Eight Days in Hell," by Camp.

31. "Battle of Peleliu 75th Anniversary Commemoration at the WWII Memorial," *WWII Memorial Friends.*

32. "The Approach to the Philippines," by Smith, chapt. XXI, p. 498.

33. "Visit Peleliu Island to Learn More About the Battle of Peleliu and Bloody Nose Ridge," by Chris Lubba, *Peleliu Dive Adventures,* 9 February 2017.

34. "WWII Battlefield Survey of Peleliu Island . . . ", by Knecht, Price, and Lindsay, p. 8.

35. Inoue interview, dtd Mar 50, in Warden ltr., *in* Western Pacific Operations," by Garand and Strobridge, p. 222.

36. "Peleliu: A Second-Generation Perspective," by Matthew Stevenson, *Military History Quaterly,* Winter 1998, *historynet.com,* in "Battle of Peleliu" *Facts and Details.*

37. "Dolphin Bay Resort and Peleliu Divers," n.d.

38. *Bloody Beaches,* by Gayle, p. 15.

39. "WWII Battlefield Survey of Peleliu Island . . .", by Knecht, Price, and Lindsay, pp. 44, 154, 182, 183, 226, 232.

40. "Preliminary Draft -- 1/91 (of Peleliu study)" (N.p.: Univ. of Hawaii, January 1991.)

41. "National Register of Historic Places Inventory—Nomination Form," item 7, p. 2.

42. Ibid., item 7, p. 3.

43. *The Assault on Peleliu,* by Hough, picture caption on p. 172.

44. Ibid., p. 136.

45. *Bloody Beaches,* by Gayle, p. 27.

46. Ibid., p. 169; "The Approach to the Philippines," by Smith, chapt. XXIV, p. 565.

Chapter 8: *Rest of the Initial Ambush*

1. President John F. Kennedy, quote from his 1962 West Point graduation speech, in "Unconventional Warfare Fundamentals."

2. *The Assault on Peleliu,* by Hough, p. 193.

3. *Home of the Heroes,* Navy Cross citation for Cpl. John O. Kincaid, Jr.

4. Ibid., Navy Cross citation for Cpl. Clarence Terrell Nicholas.

5. Ibid., Navy Cross citation for Cpl. Stanley Bernard Radomski.

6. Ibid., Navy Cross citation for PFC Charles E. Welch.

7. *The Assault on Peleliu,* by Hough, p. 132.

8. "Western Pacific Operations," by Garand and Strobridge, p. 117.

9. *Marines in the Central Solomons,* by Rentz, "USMC Historical Monograph," chapt. 4.

10. *Bloody Beaches,* by Gayle, p. 46; *With the Old Breed,* by Sledge, map on page 73.

11. *Bloody Beaches,* by Gayle, p. 13.

12. *Home of the Heroes,* Navy Cross citation for Lt.Col. Russell E. Honsowetz.

13. Ibid., Navy Cross citation for PFC Robert L. Bungard.

14. Ibid., Navy Cross citation for PFC Isadore Fessler.

15. Ibid., Navy Cross citation for Lt.Col. Robert W. Boyd.

16. Ibid., Navy Cross citation for Capt. John W. Holland.
17. Ibid., Navy Cross citation for PFC John D. Coppedge.
18. Ibid., Navy Cross citation for Cpl. Jay C. Blakely.
19. Ibid., Navy Cross citation for Sgt. James D. Miller.
20. Ibid., Navy Cross citation for PFC Dale C. Lyth.
21. Ibid., Navy Cross citation for Capt. John W. Holland.
22. Ibid., Navy Cross citation for Maj. Gordon D. Gayle.
23. Ibid., Navy Cross citation for 1stLt. Donald L. Mallory.
24. "Eight Days in Hell," by Camp, map.
25. "Western Pacific Operations," by Garand and Strobridge, p. 135.
26. *Home of the Heroes*, Navy Cross citation for Lt.Col. R.E. Honsowetz.
27. Ibid., Navy Cross citation for 2ndLt. James B. Hunter, Jr.
28. Ibid., Navy Cross citation for 1stLt. Donald L. Mallory.
29. Ibid., Navy Cross citation for 2ndLt. Edward Conroy.
30. *With the Old Breed,* by Sledge, p. 78.
31. Ibid., 79, 80.
32. "The Battle of Peleliu: Shocked Beyond Imagination," by Hemingway.
33. *Home of the Heroes*, Navy Cross citation for lstLt. Joseph A. Fournier.
34. "Eight Days in Hell," by Camp.
35. Ibid.
36. "Eight Days in Hell," by Camp.
37. "National Register of Historic Places Inventory—Nomination Form," item 8, p. 4; *The Assault on Peleliu,* by Hough, p. 63.

Chapter 9: *The Forward Slopes of Bloody Nose Ridge*

1. *Bloody Beaches,* by Gayle, p. 19.
2. *Wikipedia Encyclopedia,* s.v. "Battle of Peleliu."
3. "Personal Narrative," by Gen. O.P. Smith, Asst. Div. Cmdr., in *The Assault on Peleliu,* by Hough, p. 77.
4. "Western Pacific Operations," by Garand and Strobridge, p. 73.
5. *The Assault on Peleliu,* by Hough, p. 77.
6. Ibid.; "Western Pacific Operations," by Garand and Strobridge, p. 144.
7. *Home of the Heroes*, Navy Cross citation for Lt.Col. Russell E. Honsowetz.
8. *The Assault on Peleliu,* by Hough, p. 78.
9. Ibid., pp. 80, 81.
10. "Eight Days in Hell," by Camp.

11. "Western Pacific Operations," by Garand and Strobridge, p. 147.
12. *The Assault on Peleliu,* by Hough, p. 81.
13. "Eight Days in Hell," by Camp.
14. Ibid.
15. "The Approach to the Philippines," by Smith, pp. 533, 534.
16. *Home of the Heroes*, Navy Cross citation for Sgt. Steve Gerycz.
17. Ibid., Navy Cross citation for 1stLt. Joseph A. Fournier.
18. "Western Pacific Operations," by Garand and Strobridge, pp. 156, 157.
19. *The Assault on Peleliu,* by Hough, p. 84.
20. Ibid., p. 87.
21. "Western Pacific Operations," by Garand and Strobridge, p. 157.
22. *The Assault on Peleliu,* by Hough, p. 88.
23. *Bloody Beaches,* by Gayle, p. 26.

Chapter 10: *The Attempt to Climb Walt Ridge*

1. "The Approach to the Philippines," by Smith, pp. 533, 534.
2. "Battle of Peleliu" *Facts and Details,* n.d.
3. "Unnecessary Hell: The Battle of Peleliu," by Eric Niderost, *Warfare History Network*, 10 August 2016; *Bloody Beaches,* by Gayle, p. 42.
4. "Eight Days in Hell," by Camp.
5. *Home of the Heroes*, Navy Cross citation for Sgt. John W. Bartlett.
6. *The Assault on Peleliu,* by Hough, pp. 84-86.
7. "Unnecessary Hell: The Battle of Peleliu," by Niderost.
8. Ibid.
9. "Eight Days in Hell," by Camp.
10. "Unnecessary Hell: The Battle of Peleliu," by Niderost.
11. *Home of the Heroes*, Navy Cross citation for 2ndLt. Francis T. Burke.
12. Ibid., Navy Cross citation for Plt.Sgt. James P. McAlarnis.
13. "Unnecessary Hell: The Battle of Peleliu," by Niderost.
14. Ibid.
15. *The Assault on Peleliu,* by Hough, pp. 84-86.
16. *Home of the Heroes*, Navy Cross citation for Gy.Sgt. Warren H. Tracey.
17. *Wikipedia Encyclopedia,* s.v. "Everett P. Pope" (Medal of Honor citation).
18. *Home of the Heroes*, Silver Star citation for PFC Walter Packanik.
19. "Eight Days in Hell," by Camp.

20. Ibid.
21. Ibid.
22. "The Approach to the Philippines," by Smith, pp. 533, 534.
23. *The Assault on Peleliu,* by Hough, pp. 87, 88.
24. *Home of the Heroes*, Navy Cross citation for PFC Merle E. Manahan.
25. Ibid., Navy Cross citation for 2ndLt. Francis D. Rineer.
26. "Storming the Point at Peleliu," by Camp.
27. Ibid.
28. "Battle of Peleliu Facts," by Mack Dean, *WorldWar2Facts,* 2 June 2020.
29. "National Register of Historic Places Inventory—Nomination Form," item 7, p. 3.
30. Ibid., p. 3.
31. Ibid., item 8, p. 5.
32. *The Marines,* ed. (B.Gen.) Edwin Howard Simmons and J. Robert Moskin (Quantico, VA: Marine Corps Heritage Foundation, 1998), p. 67.

Chapter 11: *Degree of Tunneling in the Southern Massif*

1. "WWII Battlefield Survey of Peleliu Island . . .", by Knecht, Price, and Lindsay, p. 40.
2. Ibid., p. 236.
3. Ibid., p. 239.
4. "Peleliu 1944: The Archaeology of a South Pacific D-Day," Price and Knecht, p. 33.
5. "WWII Battlefield Survey of Peleliu Island . . .", by Knecht, Price, and Lindsay, p. 11.
6. Ibid., p. 230.
7. Ibid.
8. Ibid., pp. 1-276.
9. Ibid., p. 135.
10. *Bloody Beaches,* by Gayle, p. 3.
11. "WWII Battlefield Survey of Peleliu Island . . .", by Knecht, Price, and Lindsay, p. 100.
12. Ibid., p. 85.
13. Ibid., p. 33; *The Assault on Peleliu,* by Hough, pp. 80, 81.
14. *Bloody Beaches,* by Gayle, p. 20.
15. *Wikipedia Encyclopedia,* s.v. "John Dury New" (Medal of Honor citation).
16. Ibid., s.v. "Charles Howard Roan" (Medal of Honor citation).
17. *Home of the Heroes*, Navy Cross citation for PFC Charles J. Pottersnak.

18. Ibid., Navy Cross citations for 2ndLt. Francis T. Burke, Plt. Sgt. James P. McAlarnis, and PFC Merle E. Manahan.
19. Ibid., Navy Cross citation for 2ndLt. Francis D. Rineer.
20. Ibid., Navy Cross citation for PFC Boone T. Harris.
21. Ibid., Navy Cross citation for PFC Robert L. Bungard.
22. Ibid., Navy Cross citation for Cpl. John O. Kincaid, Jr.
23. *The Assault on Peleliu,* by Hough, p. 78.
24. "WWII Battlefield Survey of Peleliu Island . . . ", by Knecht, Price, and Lindsay, p. 203.
25. *Home of the Heroes*, Navy Cross citation for Cpl. Clarence Terrell Nicholas.
26. Ibid., Navy Cross citation for Sgt. Joseph D. Sheehan.

Chapter 12: *Foe Who Kept Popping Up Behind U.S. Lines*

1. *Wikipedia Encyclopedia,* s.v. "Carlton R. Rouh' (Medal of Honor citation).
2. *Home of the Heroes,* Navy Cross citations for Cpl Stanley Bernard Radomski and PFC Charles E. Welch.
3. *The Assault on Peleliu,* by Hough, pp. 136, 193-196.
4. "Reflections on Iwo Jima: The Japanese Tunnels," interview of Col. Dave Severance USMC (Ret.), CO of E/2/28, *Marines TV,* Youtube video, 26 March 2011.
5. *The Assault on Peleliu,* by Hough, p. 193.
6. Peter Antill, "Peleliu, battle for (Operation Stalemate II)— The Pacific War's Forgotten Battle, September-November 1944," in "Battle of Peleliu," *HistoryofWar.org.*
7. *Bloody Beaches,* by Gayle, p. 6; *The Assault on Peleliu,* by Hough, p. 169; "The Approach to the Philippines," by Smith, chapt. XXIV, p. 565.
8. *Mao's Generals Remember Korea,* p. 154.
9. *Bloody Beaches,* by Gayle, pp, 27, 28.
10. "WWII Battlefield Survey of Peleliu Island . . . ," by Knecht, Price, and Lindsay, p. 85.
11. Ibid.
12. *The Assault on Peleliu,* by Hough, p. 132.
13. Ibid., p. 132, footnote 27.
14. "The Approach to the Philippines," by Smith, chapt. XXIV, p. 545.
15. *Bloody Beaches,* by Gayle, p. 43.
16. "WWII Battlefield Survey of Peleliu Island . . . ," by Knecht, Price, and Lindsay, p. 135.
17. "The Approach to the Philippines," by Smith, chapt. XXIV, pp. 541, 566.
18. Ibid., pp. 567, 568.

Chapter 13: *Initial Access to Top of the Umurbrogol*

1. "National Register of Historic Places Inventory—Nomination Form," item 8, p. 5.
2. Antill, "Peleliu, battle for (Operation Stalemate II) . . . ," in "Battle of Peleliu." *HistoryofWar.org*.
3. "National Register of Historic Places Inventory—Nomination Form," item 8, p. 6.
4. "The Approach to the Philippines," by Smith, pp. 554, 555.
5. Ibid., pp. 556, 557.
6. Ibid., p. 558.
7. Ibid.
8. Ibid., p. 559.
9. Ibid.
10. "Western Pacific Operations," by Garand and Strobridge, pp. 240, 241.
11. Ibid.
12. "The Approach to the Philippines," by Smith, chapt. XXIV, p. 566.
13. Ibid., pp. 566, 567.

Chapter 14: *Scramble for the Pocket*

1. "Peleliu 1944: The Archaeology of a South Pacific D-Day," by Price and Knecht, p. 10.
2. "WWII Battlefield Survey of Peleliu Island . . . ," by Knecht, Price, and Lindsay, pp. 134, 187, 195, 199; *Bloody Beaches,* by Gayle, pp. 27, 28.
3. Antill, "Peleliu, battle for (Operation Stalemate II) . . . ," in "Battle of Peleliu," *HistoryofWar.org*.
4. "National Register of Historic Places Inventory—Nomination Form," item 8, p. 7.
5. "The Battle of Peleliu: Shocked Beyond Imagination," by Hemingway.
6. "WWII Battlefield Survey of Peleliu Island . . . ," by Knecht, Price, and Lindsay, p. 138.
7. "The Approach to the Philippines," by Smith, chapt. XXIV, p. 565.
8. Ibid., p. 559.
9. Ibid., p. 568.
10. Ibid., p. 571.
11. Ibid., p. 572; "The Battle of Peleliu: Shocked Beyond Imagination," by Hemingway.

12. "Peleliu 1944: The Archaeology of a South Pacific D-Day," by Price and Knecht, p. 33.

Chapter 15: *Peleliu Lessons*

1. "Battle of Peleliu" *Facts and Details,* n.d., as retrieved in December 2020 from the following url: http://factsanddetails.com/asian/ca67/sub428/entry-5336.html.
2. "Map May Hold Key," by David. S. Cloud, *Los Angeles Times,* 3 May 2015.
3. "Closing In: Marines in the Seizure of Iwo Jima," by Col. Joseph H. Alexander, *Marines in World War II Commemorative Series* (Washington, D.C.: Hist. and Museums Div., HQMC, 1994), p. 37.
4. "Operation Stalemate II: The Battle of Peleliu 15 September–27 November 1944" (N.p.: Naval Hist. and Heritage Cmd., n.d.).
5. "Peleliu 1944: The Archaeology of a South Pacific D-Day," Price and Knecht, p. 15.
6. "Visit Peleliu Island to Learn More About the Battle of Peleliu and Bloody Nose Ridge," by Lubba.
7. "Place of the Loyal Samurai," by Weiss.
8. "WWII Battlefield Survey of Peleliu Island . . . ", by Knecht, Price, and Lindsay, p. 225; "Peleliu 1944: The Archaeology of a South Pacific D-Day," Price and Knecht, p. 33.
9. "Battle of Peleliu" *Facts and Details,* n.d., as retrieved in December 2020 from the following url: http://factsanddetails.com/asian/ca67/sub428/entry-5336.html.
10. *Encyclopedia Britannica* on line, s.v. "Occam's Razor."
11. "Fighting on Guadalcanal," U.S.A. War Office pamphlet (Washington, D.C.: U.S. Govt. Printing Office, 1942), pp, 66, 67.
12. "WWII Battlefield Survey of Peleliu Island . . . ", by Knecht, Price, and Lindsay, photos in figures 6.1 and 6.2, pp. 134, 135; "Preliminary Draft -- 1/91 (of Peleliu study)" (N.p.: Univ. of Hawaii, January 1991).
13. *With the Old Breed,* by Sledge, pp. 63, 79.
14. "Western Pacific Operations," by Garand and Strobridge, p. 39.
15. *The Assault on Peleliu,* by Hough, p. 16.
16. Michael Lee Lanning and Dan Cragg, *Inside the VC and the NVA: The Real Story of North Vietnam's Armed Forces* (New York: Ivy Books, 1992), p. 211.
17. *The Assault on Peleliu,* by Hough, p. 181.
18. *Wikipedia Encyclopedia, s.v.* "Spit (Land Form)."
19. "Preliminary Draft -- 1/91 (of Peleliu study)," (N.p.: Univ. of Hawaii, January 1991).

20. "Storming the Point at Peleliu," by Camp; *Wikipedia Encyclopedia,* s.v. "Battle of Peleliu."
21. *The Assault on Peleliu,* by Hough, p. 196.
22. Ibid., p. 135.
23. "The Battle of Biak: 'A Terrifying Glimpse Into the Soul of Mankind'," by Johnson; "The Approach to the Philippines," by Smith, pp. 391, 392, and footnotes 41, 47.
24. *The Assault on Peleliu,* by Hough, p. 194.
25. Ibid., p. 193.
26. Ibid., pp. 181, 182.
27. *The Tiger's Way,* by Poole, p. 178.
28. *Home of the Heroes*, Navy Cross citation for PFC George Lilja.
29. *Super-Squad,* by Poole, chapt. 11.
30. Ibid., Navy Cross citation for PFC Ted. J. Carlton.

Chapter 16: *What Made the Taking of Peleliu Important*

1. "Army Is Spending Half a Billion to Train Soldiers to Fight Underground," by Matthew Cox, *military.com,* 24 June 2018.
2. "The Battle of Peleliu: Shocked Beyond Imagination," by Hemingway.
3. "The Approach to the Philippines," by Smith, chapt. XXI, map VI.
4. *Wikipedia Encyclopedia*, s.v. "Battle of Bint Jbeil."
5. *The Iwo Alamo,* by Poole, chapt. 13.
6. "Western Pacific Operations," by Garand and Strobridge, pp. 41, 112.
7. "Eight Days in Hell," by Camp.
8. *The Iwo Alamo,* by Poole, chapts. 10 and 11, p. 185.
9. *Gung Ho,* by Poole, chapt. 10.
10. "The Final Campaign: Marines in the Victory on Okinawa," by Col. Joseph Alexander, *Marines in World War II Commemorative Series,* p. 28.
11. *Super-Squad,* by Poole, epilogue.
12. Eugene Sledge, as quoted in *The War,* by Burns and Novick, part 6.
13. "Battle of Leyte Gulf: Facts, Casualties, & Significance," *Encyclopedia Britannica,* 16 October 2020.
14. "The Final Campaign: Marines in the Victory on Okinawa," by Alexander, pp. 1, 2.
15. *The Assault on Peleliu,* by Hough, p. 7.
16. *Wikipedia Encyclopedia,* s.v. "Greater East Asia Co-Prosperity Sphere," Wikimedia Commons global image (colored portions).

17. Eric Hammel, *Fire in the Streets: The Battle for Hue, Tet 1968.* (Pacifica, CA: Pacifica Press, 1991), pp. 304, 305.

18. Ibid. p. 274.

19. *The Iwo Alamo,* by Poole, chapt. 13.

20. Thomas M. Huber, "Japan's Battle for Okinawa, April - June 1945," *Leavenworth Papers No. 18* (Ft. Leavenworth, KS: Combat Studies Inst., U.S. Army's Cmd. & Gen. Staff College, 1990), chapt. 3.

21. "John 15:13," *Christian Bible,* King James Version, Bible Gateway.

Glossary

ACMS	Anti-Cruise Missile System	Equip. for shooting down cruise missiles
ADIZ	Air Defense Identification Zone	No-fly area.
A.k.a.	Also known as.	An alias.
AP	Associated Press	U.S. news agency.
CNN	Cable News Network	U.S. news agency.
CO	Commanding Officer	Senior unit leader.
CP	Command post.	Location of unit CO
D-Day	Debarkation-Day	Start of important invasion or operation
DoD	Department of Defense	Pentagon agency.
DUKW	D means 1942, U means utility (amphibious), K means all-wheel drive, W means two powered rear axles.	Type of military wheeled amphibious landing craft.
E-1	U.S. military pay grade	Marine Private's enlistment status.
E-2	U.S. military pay grade	Marine PFC's status.
4GW	Fourth Generation Warfare	War waged in four arenas at once—martial, religious or psychological, economic, political

FPL	Final Protective Line	Interlocking bands of defensive MG fire.
GI	Government Issue	Colloquial term for low-ranking member of U.S. Armed Forces
HQMC	Headquarters Marine Corps	Top echelon of USMC.
IJA	Imperial Japanese Army	Ground forces of Japan.
JCS	Joint Chiefs of Staff	Heads of all U.S. service branches
LACM	Land Attack Cruise Missile	Cruise missile for firing at ship from the shore.
LVT	Landing Vehicle Tracked	Amphibian tractor
LVTA	LVT Armored	LVT with an artillery piece or flame thrower
MG	Machinegun	Automatic rifle.
MLR	Main Line of Resistance	Linear formation of defensive positions.
MoH	Medal of Honor	Highest U.S. award for valor.
MSN	*Microsoft News*	U.S. news agency.
MW	Maneuver Warfare	A style of fighting that depends more on maneuver than firepower. Often equated with 3GW.
NATO	North Atlantic Treaty Organization	Western military alliance.

NCO	Non-Commission Officer	Enlisted leader.
NLG	Northern Landing Group	Part of U.S. New Georgia contingent
NPR	*National Public Radio*	Educational radio.
NVA	North Vietnamese Army	Hanoi-directed ground forces.
OP	Observation Post	Forward sentry site.
PFC	Private First Class	Second-echelon enlisted rank.
PLA	People's Liberation Army	PRC's armed forces.
PLAN	People's Liberation Army Navy	Naval branch of PLA
POW	Prisoner of War	Captured combatant.
PRC	People's Republic of China	Mainland China.
RCT	Regimental Combat Team	Reinforced infantry regiment.
RPG	Rocket-Propelled Grenade	Personal antitank weapon.
SAMS	Surface-to-Air Missiles	Weapons for shooting down airplanes from the ground.
2GW	2nd Generation Warfare	Focus is on killing enemy soldiers; same as traditional U.S. style (attrition)
SNCO	Staff Non-Commissioned Officer	Higher-echelon enlisted leader.
SOP	Standard Operating Procedure	Long-established methods for doing things in the military.

3GW	3rd Generation Warfare	Bypasses manpower concentrations of foe to more easily get at his command and logistics networks.
TNT	Trinitrotoluene	Explosive used in WWII satchel charge.
U.N.	United Nations	Worldwide alliance of countries.
U.S.	United States	America.
USMC	U.S. Marine Corps	America's amphibious force in readiness.
UW	Unconventional Warfare	Attacking through infiltration and defending through ambushes; part of 3GW.
UXO	Unexploded Ordnance	Live shells and bombs
VC	Viet Cong	Enemy guerrillas in Vietnam
WWI	World War I	First global conflict.
WWII	World War II	Second global conflict

Bibliography

U.S. Government Publications, News Releases, and Videos

"The Approach to the Philippines." By Robert Ross Smith. *U.S. Army in WWII Series*. Washington, D.C.: Center of Military History, 1996. As retrieved in September 2020 from these urls: https://history.army.mil/html/books/005/5-8-1/cmhPub_5-8-1.pdf; and https://www.ibiblio.org/hyperwar/USA/USA-P-Approach/index.html.

The Assault on Peleliu. By Maj. Frank R. Hough USMCR. Washington, D.C.: Historical Division, Headquarters Marine Corps, 1950.

Bloody Beaches: The Marines at Peleliu. By B.Gen. Gordon D. Gayle USMC (Ret.). *Marines in WWII Commemorative Series*. Washington, D.C.: Marine Corps Historical Center, 1996. As retrieved in September 2020 from these urls: Bloody Beaches The Marines At Peleliu PCN 19000313701.pdf; and https://www.ibiblio.org/hyperwar/USMC/USMC-C-Peleliu/index.html.

"Closing In: Marines in the Seizure of Iwo Jima." By Col. Joseph H. Alexander. *Marines in World War II Commemorative Series*. Washington, D.C.: History and Museums Division, Headquarters Marine Corps, 1994. As retrieved in January 2020 from the following urls: https://ia902707.us.archive.org/20/items/ClosingIn/ClosingIn.pdf; https://www.nps.gov/parkhistory/online_books/npswapa/extcontent/usmc/pcn-190-003131-00/pcn-190-003131-00/index.htm.

"Fighting on Guadalcanal." U.S.A. War Office pamphlet. Washington, D.C.: U.S. Government Printing Office, 1942.

"The Final Campaign: Marines in the Victory on Okinawa." By Col. Joseph Alexander. *Marines in World War II Commemorative Series*. Washington, D.C.: History and Museums Division, Headquarters Marine Corps, 1996. As retrieved in September 2020 from the following url: https://www.ibiblio.org/hyperwar/USMC/USMC-C-Okinawa/index.html.

"From Makin to Bougainville: Marine Raiders in the Pacific War." By Maj. Jon T. Hoffman. *Marines in WWII Commemorative Series*. Washington, D.C.: Marine Corps Historical Center, 1995. As retrieved in September 2020 from this url: https://www.ibiblio.org/hyperwar/USMC/USMC-C-Raiders/index.html.

Handbook on Japanese Military Forces. TM-E 30-480. Washington, D.C.: U.S. War Department, 1944. Reprint, Baton Rouge, LA: LSU Press, 1991.

"History of the Biak Operation 15-27 June 1944." U.S. 41st Army, 1944. Armed Forces Staff College Library copy. As retrieved in September 2020 from the following url: https://apps.dtic.mil/dtic/tr/fulltext/u2/a550758.pdf.

Huber, Thomas M. "Japan's Battle for Okinawa, April - June 1945." *Leavenworth Papers No. 18*. Ft. Leavenworth, KS: Combat Studies Institute, U.S. Army's Command and General Staff College, 1990.

"Isolation of Rabaul." By Henry I. Shaw and Major Douglas T. Kane. *History of U.S. Marine Corps Operations in World War II Series*. Volume II. Washington, D.C.: Historical Branch, Headquarters Marine Corps, 1963. As retrieved from the following url: www.ibiblio.org/hyperwar/USMC/II/index.html#contents.

Iwo Jima: Amphibious Epic. Lt.Col. Whitman S. Bartley. Washington, D.C.: Historical Branch, Headquarters Marine Corps, 1954.

"Marines in the Central Solomons." By Maj. John N. Rentz, *U.S. Marine Corps Historical Monograph*. Washington, D.C.: History Branch, Headquarters Marine Corps, 1952.

"National Register of Historic Places Inventory—Nomination Form (for Peleliu Battlefield, Palau Islands)." National Park Service. N.p.: United States Department of the Interior, n.d. As retrieved in November 2020 from the following url: https://npgallery.nps.gov/GetAsset/b8f8b24e-253b-4111-937a-e3515b7f9ffd.

"Nomonhan: Japanese—Soviet Tactical Combat, 1939." By Edward J. Drea. *Leavenworth Papers No. 2*. Ft. Leavenworth, KS: Combat Studies Institute, U.S. Army Command and General Staff College, 1981.

North Korea Handbook. PC-2600-6421-94. Washington, D.C.: Department of Defense, 1994.

"Operation Stalemate II: The Battle of Peleliu 15 September–27 November 1944." N.p.: Naval History and Heritage Command, n.d. As retrieved in December 2020 from the following url: https://www.history.navy.mil/browse-by-topic/wars-conflicts-and-operations/world-war-ii/1944/peleliu.html.

"Preliminary Draft -- 1/91 (of Peleliu study)." Botany Department. N.p: University of Hawaii, January 1991. (May contain some of the same content as the "Peleliu Report—University of Hawaii Botany Department," by D. Colt Denfeld, in 1982, for the Historic Preservation Office, Trust Territory of the Pacific Islands.) As retrieved in November 2020 from the following url: http://www.botany.hawaii.edu/basch/uhnpscesu/htms/peleliu/pelbody.htm.

"Reflections on Iwo Jima: The Japanese Tunnels." Interview of Col. Dave Severance USMC (Ret.), CO of E/2/28. *Marines TV.* Youtube video, 26 March 2011.

"Western Pacific Operations." *History of Marine Corps Operations in World War II.* Volume IV. Part III. By George W. Garand and Truman R. Strobridge. N.p.: Historical Branch, G-3 Division, Headquarters Marine Corps, 1971. As retrieved in October 2020 from the following url: ibiblio.org/hyperwar/USMC/IV/index.html.

"WWII Battlefield Survey of Peleliu Island Peleliu State, Republic of Palau," Submitted for the Peleliu War Historical Society by Rick Knecht, Neil Price, and Gavin Lindsay. Department of Archeology, University of Aberdeen, England. Washington, D.C.: National Park Service American Battlefield Protection Program, 2012. As retrieved in January 2021 from the following url: https://www.researchgate.net/profile/Rick_Knecht/publication/346648924_WWII_Battlefield_Survey_of_Peleliu_Island_Republic_of_Palau_Final_Report_July_2012/links/5fcb7f7e299bf188d4f60abc/WWII-Battlefield-Survey-of-Peleliu-Island-Republic-of-Palau-Final-Report-July-2012.pdf.

Civilian Publications

Analytical Studies and Websites

Association of U.S. Marine Raiders' website. As retrieved from this url: http://www.usmarineraiders.org/.

Bible Gateway. From its website, biblegateway.com.

Chesty: The Story of Lieutenant General Lewis B. Puller. By Col. Jon T. Hoffman USMCR. New York: Random House, 2002.

Encyclopedia Britannica. From its website, https://www.britannica.com.

Gertz, Bill. *The China Threat: How the People's Republic Targets America.* N.p.: Regnery Publishing, 2000.

Goulden, Joseph C. *Korea: The Untold Story of the War.* New York: Times Books, 1982.

Guardian Joe: How Less Force Helps the Warrior. By H. John Poole. Emerald Isle, NC: Posterity Press, 2018.

Gudmundsson, Bruce I. *Stormtroop Tactics—Innovation in the German Army 1914-1918.* New York: Praeger, 1989.

Gung Ho: The Corps' Most Progressive Tradition. By H. John Poole. Emerald Isle, NC: Posterity Press, 2012.

Hall of Valor Project. From its website, https://valor.militarytimes.com.

Hammel, Eric. *Fire in the Streets: The Battle for Hue,* Tet 1968. Pacifica, CA: Pacifica Press, 1991.

Home of the Heroes. From its website, https://homeofheroes.com.

The Iwo Alamo. By H. John Poole. Emerald Isle, NC: Posterity Press, 2020.

"Japanese Defenses and Fortifications Tawara, Iwo, Jima, and Okinawa 1943-1945." By J. David Rogers and Karl F. Hasserman Chair. University of Missouri at Rolla, n.d. As retrieved in February 2020 from the following url: https://web.mst.edu/~rogersda/umrcourses/ge342/Japanese%20Island%20Defenses%201943-45.pdf.

Lanning, Michael Lee and Dan Cragg. *Inside the VC and the NVA: The Real Story of North Vietnam's Armed Forces.* New York: Ivy Books, 1992.

Mao's Generals Remember Korea. Translated and edited by Xiaobing Li, Allan R. Millett, and Bin Yu. Lawrence, KS: University Press of Kansas, 2001.

The Marines. Edited by (B.Gen.) Edwin Howard Simmons and J. Robert Moskin, Quantico, VA: Marine Corps Heritage Foundation, 1998.

Menges, Constantine. *China: The Gathering Threat.* Nashville, TN: Nelson Current, 2005.

North Korea's Hidden Assets. By H. John Poole. Emerald Isle, NC: Posterity Press, 2018.

NPR News on line. From the NPR website, http://www.npr.org.

The Pacific War Online Encyclopedia. From its website, http://pwencycl.kgbudge.com.

Shaw, John and the editors of Time-Life Books. *Red Army Resurgent.* Chicago: Time-Life Books, 1979.

Phantom Soldier: The Enemy's Answer to U.S. Firepower. By H. John Poole. Emerald Isle, NC: Posterity Press, 2001.

Strategic Rifleman: Key to More Moral Warfare. By H. John Poole. Emerald Isle, NC: Posterity Press, 2014.

Super-Squad: The Now Missing Component. By H. John Poole. Emerald Isle, NC: Posterity Press, January 2020.

The Tiger's Way: A U.S. Private's Best Chance at Survival. By H. John Poole. Emerald Isle, NC: Posterity Press, 2003.

Warner, Philip. "Japanese Army of World War II." *Men-at-Arms Series.* Color plates by Michael Youens. London: Osprey Publications Limited, 1972.

With the Old Breed: At Peleliu and Okinawa. By E.B. Sledge. New York: Presidio Press, a division of Random House, 2007.

Yahara, Col. Hiromichi. *The Battle for Okinawa.* Translated Roger Pineau and Masatoshi Uehara. New York: John Wiley & Sons, 1995.

Videotapes, Movies, DVDs, TV Programs, Slide Shows, CDs, Illustrations

Sands of Iwo Jima. Hollywood: Republic Pictures, 1988. Videocassette.

The War. Produced by Ken Burns and Lynn Novick. N.p.: Sony Legacy, 2007. Film documentary. Part 6. Public Broadcasting Service, 12 September 2020.

Newspaper, Magazine, and Website Articles

"Army Is Spending Half a Billion to Train Soldiers to Fight Underground." By Matthew Cox. *Military.com,* 24 June 2018. As retrieved in December 2020 from the following url: military.com, https://www.military.com/daily-news/2018/06/24/army-spending-half-billion-train-troops-fight-underground.html.

"The Battle of Biak: 'A Terrifying Glimpse Into the Soul of Mankind.'" By David Alan Johnson. *Warfare History Network,* n.d. As retrieved in September 2020 from the following url: https://warfarehistorynetwork.com/2018/12/30/the-battle-of-biak-a-terrifying-glimpse-into-the-soul-of-mankind/.

"Battle of Leyte Gulf | Facts, Casualties, & Significance." *Encyclopedia Britannica,* 16 October 2020. As retrieved in November 2020 from the following url: https://www.britannica.com/event/Battle-of-Leyte-Gulf.

"Battle of Peleliu." *Facts and Details.* As retrieved in November 2020 from the following url: http://factsanddetails.com/asian/ca67/sub428/entry-5336.html.

"Battle of Peleliu." *HistoryofWar.org.* As retrieved in January 2023 from this url: http://www.historyofwar.org/articles/battles_peleliu.html#pocket.

"Battle of Peleliu Facts." By Mack Dean. *WorldWar2Facts,* 2 June 2020. As retrieved in November 2020 from the following url: http://www.worldwar2facts.org/battle-of-peleliu-facts.html.

"Battle of Peleliu 75th Anniversary Commemoration at the WWII. Memorial." *WWII Memorial Friends.* As retrieved in November 2020 from the following url: https://wwiimemorialfriends.networkforgood.com/events/15453-battle-of-peleliu-75th-anniversary-commemoration-at-the-wwii-memorial.

"The Battle of Peleliu: Shocked Beyond Imagination." By Al Hemingway, *Warfare History Network,* 4 January 2019. As retrieved in November 2020 from the following url: https://warfarehistorynetwork.com/2019/01/04/the-battle-of-peleliu-shocked-beyond-imagination/.

"China Defends Land Reclamation in the Spratly Islands." *Military Times* on line, 24 November 2014.

"China's Military Rise: The Dragon's New Teeth." *The Economist,* 7 April 2012.

"China's Newest Maritime Dispute." By Zachary Keck. *The Diplomat* on line, 20 March 2014.

"China Stations Combat Aircraft on South China Sea Island." By Franz-Stefan Gady. *The Diplomat* on line, 10 November 2015.

"China to soon build air, naval bases in Scarborough Shoal, Carpio warns." By CNN Staff. *CNN Philippines* website, 9 June 2020. As retrieved in September 2020 from the following url: https://www.cnnphilippines.com/news/2020/6/9/China-build-air-naval-bases-Scarborough-Shoal.html.

"Dolphin Bay Resort and Peleliu Divers." As retrieved in November 2020 from the following url: http://dolphinbay-resort-peleliu.com/peleliu_island.html.

"Eight Days in Hell." By Colonel Richard D. Camp, USMC (Ret.) *Naval History Magazine.* Volume 34, number 5, 19 October 2019. As retrieved in August 2020 from this url: https://www.usni.org/magazines/naval-history-magazine/2019/october/eight-days-hell.

"How the Scarborough Shoal came back to haunt China-Philippines relations." By Richard Heydarian. *South China Morning Post* on line, 23 June 2018. As retrieved in February 2021 from the following url: https://www.scmp.com/news/china/diplomacy-defence/article/2151923/how-scarborough-shoal-came-back-haunt-china-philippines.

"How Uninhabited Islands Soured Japan-China Ties." *BBC News* on line, 10 November 2014.

"Inside the International Contest over the Most Important Waterway in the World." By Hannah Beech. *Time Magazine* on line, 6 June 2016.

"Japan WW2 soldier who refused to surrender Hiroo Onoda dies." *BBC News* on line, 17 January 2014. As retrieved in September 2020, from the following url: https://www.bbc.com/news/world-asia-25772192.

Kennedy, Pesident John F. Quote from his 1962 West Point graduation speech. In "Unconventional Warfare Fundamentals," at *The Irregular Warrior* website.

"Map May Hold Key." By David. S. Cloud. *Los Angeles Times,* 3 May 2015. As retrieved in December 2020 from the following url: https://www.latimes.com/world/asia/la-fg-wwii-peleliu-20150503-story.html.

"The Meaning of China's Defeat in the South China Sea." By Hannah Beech. *Time Magazine* on line, 25 July 2016.

"Military Build-Up in the South China Sea." By Eric Grossman, Rand Corporation website, December 2019. As retrieved in September 2020 from the following url: www.rand.org/content/dam/rand/pubs/external_publications/EP60000/EP68058/RAND_EP68058.pdf.

The Pacific War Online Encyclopedia. From its website, http://pwencycl.kgbudge.com.

"Peleliu 1944: The Archaeology of a South Pacific D-Day." By Neil Price and Rick Knecht. Department of Archaeology. University of Aberdeen, UK. *Journal of Conflict Archaeology.* Volume 7, number 1, January, 2012.

"Place of the Loyal Samurai." By Daniel Weiss. *Archaeology Magazine,* July/August 2019. As retrieved in September 2020 from the following url: https://www.archaeology.org/issues/344-1907/features/7712-peleliu-wwii-pacific-battle.

"Report: China Has Reclaimed 3,200 Acres in South China Sea." By Lolita Baldor (Associated Press). *MSN News* on line, 14 May 2016. As retrieved in May 2016 from this url: http://www.msn.com/en-us/news/world/report-china-has-reclaimed-3200-acres-in-south-china-sea/ar-BBt1ccE?li=BBnb7Kz.

"Storming the Point at Peleliu." By Dick Camp. *Warfare History Network,* n.d. As retrieved in August 2020 from the following url: https://warfarehistorynetwork.com/2018/12/23/storming-the-point-at-peleliu/.

"Two U.S. carrier groups conduct exercises in South China Sea." By Se Young Lee. *Reuters* on line, 9 February 2021. As retrieved in February 2021 from the following url: https://www.reuters.com/article/us-southchinasea-usa-carriers/two-u-s-carrier-groups-conduct-exercises-in-south-china-sea-idUSKBN2A90I5?il=0.

"Unnecessary Hell: The Battle of Peleliu." By Eric Niderost. *Warfare History Network,* 10 August 2016. As retrieved in November 2020 from the following url: https://warfarehistorynetwork.com/2016/08/10/unnecessary-hell-the-battle-of-peleliu/.

"Up to the Task." By Audrey McAvoy (Associated Press). *Jacksonville Daily News* (NC), 17 January 2016.

"U.S. Navy Sends Destroyer by Man-Made Islands in Challenge to China's Claims." By David Larter. *Navy Times* on line, 27 October 2015.

"U.S. Warship Sails Near Chinese-Controlled S. China Sea Islands." By Reuters Wire Service. *U.S. News & World Report* on line, 5 February 2021. As retrieved in February 2021 from the following url: https://www.usnews.com/news/world/articles/2021-02-05/us-warship-sails-near-chinese-islands-in-south-china-sea.

"Visit Peleliu Island to Learn More About the Battle of Peleliu and Bloody Nose Ridge." By Chris Lubba. Peleliu Dive Adventures website, 9 February 2017. As retrieved in November 2020 from the following url: https://palaudiveadventures.com/peleliu-island/.

About the Author

After 28 years of commissioned and noncommissioned infantry service, John Poole retired from the United States Marine Corps in April 1993. While on active duty, he studied small-unit tactics for nine years: (1) six months at the Basic School in Quantico (1966); (2) seven months as a rifle platoon commander in Vietnam (1966-67); (3) three months as a rifle company commander at Camp Pendleton (1967); (4) five months as a regimental headquarters company (and camp) commander in Vietnam (1968); (5) eight months as a rifle company commander in Vietnam (1968-69); (6) five and a half years as an instructor with the Advanced Infantry Training Company (AITC) at Camp Lejeune (1986-92); and (7) one year as the Staff Noncommissioned Officer in Charge (SNCOIC) of the 3rd Marine Division Combat Squad Leaders Course (CSLC) on Okinawa (1992-93).

While at AITC, he developed, taught, and refined courses on maneuver warfare, land navigation, fire support coordination, call for fire, adjust fire, close air support, M203 grenade launcher, movement to contact, daylight attack, night attack, infiltration, defense, offensive Military Operations in Urban Terrain (MOUT), defensive MOUT, Nuclear/Biological/Chemical (NBC) defense, and leadership. While at CSLC, he further refined the same periods of instruction and developed others on patrolling.

He has completed all of the correspondence school requirements for the Marine Corps Command and Staff College, Naval War College (1,000-hour curriculum), and Marine Corps Warfighting Skills Program. He is a graduate of the Camp Lejeune Instructional Management Course, the 2nd Marine Division Skill Leaders in Advanced Marksmanship (SLAM) Course, and the East-Coast School of Infantry Platoon Sergeants' Course.

In the 25 years since retirement, Poole has researched the small-unit tactics of other nations and written 18 other books: (1) *The Last Hundred Yards,* a squad combat study based on the consensus opinions of 1,200 formal school NCOs and casualty statistics of AITC and CSLC field trials; (2) *One More Bridge to Cross,* a treatise on enemy proficiency at short range and how to match it; (3) *Phantom Soldier,* an in-depth look at the highly deceptive Asian style of war; (4) *The Tiger's Way,* the fighting styles of Eastern fireteams and soldiers; (5) *Tactics of the Crescent Moon,* insurgent procedures in Palestine, Chechnya, Afghanistan, and Iraq; (6) *Militant Tricks,* how Muslim insurgents fight in an urban environment; (7) *Terrorist Trail,* tracing the jihadists in Iraq back to their home countries; (8) *Dragon Days,* an unconventional warfare technique manual; (9) *Tequila Junction,* how to fight narco-guerrillas; (10) *Homeland Siege,* countering the 4GW assault on America by a foreign power's organized-crime proxies; (11) *Expeditionary Eagles,* how to outmaneuver the Taliban; (12) *Global Warrior,* forestalling WWIII through tiny-contingent activity; (13) *Gung Ho,* how supporting arms are not needed to take strongpoint matrices; (14) *Strategic Rifleman,* how to produce tiny semi-autonomous combat elements; (15) *Afrique,* with Africa as a model how U.S. intelligence is being gathered incorrectly for certain adversaries; (16) *Sinoland,* how the U.S. has been under 4GW attack from China since 9/11; (17) *Guardian Joe,* how—through more skilled infantry squads—the Pentagon can use less force to more easily win a modern war; (18) *North Korea's Hidden Assets,* 13 little-known wartime capabilities of what has been erroneously called the Hermit Kingdom; (19) *Super-Squad: The Now Missing Component,* how the current state of the art for small-unit tactics was collectively achieved by the big Eastern armies late in WWII; and (20) *The Iwo Alamo,* how machinegun-hunting scouts saved the U.S. divisions.

Since retirement, Poole has traveled extensively in both Communist and Islamist worlds to study the infantry tactics of America's enemies. He has done research in Mainland China (twice), North Korea, Vietnam, Cambodia, Pakistan (twice), Bangladesh, Malaysia, Iran, Lebanon, Turkey, the Emirates, Egypt, Sudan, Tanzania, Zambia, South Africa, Morocco, and Venezuela. Over the course of his lifetime, he has visited scores of other nations on all five continents.

As of April 2014, John Poole had conducted multi-day training sessions (on advanced squad tactics) at 41 (mostly Marine) battalions, nine Marine schools, and seven special-operations units from all four U.S. service branches.

Between early tours in the Marine Corps (from 1969 to 1971), he served as a criminal investigator with the Illinois Bureau of Investigation (IBI). After attending the State Police Academy for several months in Springfield, he was assigned to the IBI's Chicago office. There, he worked mostly on general criminal and drug cases.

Poole is one of the very few Americans with an extensive enough background to know how to fix the ongoing tactical shortfall within the U.S. military. After some 60 multi-day training sessions at different active-duty units, he has developed, tested, and refined its solution (through a "bottom-up" training supplement). He has also become adept at the now virtually lost art of field training light infantrymen (those who are able do well on their own in the absence of orders).

Name Index

O

P

Q

No entries.

R

S

T

U

No entries.

V

W

X

Y

Z

No entries.